The Object [...] very close t[...] even banal—[...] a rich histo[...] science, and popular mythology. Filled with fascinating details and conveyed in sharp, accessible prose, the books make the everyday world come to life. Be warned: once you've read a few of these, you'll start walking around your house, picking up random objects, and musing aloud: 'I wonder what the story is behind this thing?'"

Steven Johnson, author of *Where Good Ideas Come From* and *How We Got to Now*

Object Lessons describes themselves as 'short, beautiful books,' and to that, I'll say, amen. . . . If you read enough Object Lessons books, you'll fill your head with plenty of trivia to amaze and annoy your friends and loved ones—caution recommended on pontificating on the objects surrounding you. More importantly, though . . . they inspire us to take a second look at parts of the everyday that we've taken for granted. These are not so much lessons about the objects themselves, but opportunities for self-reflection and storytelling. They remind us that we are surrounded by a wondrous world, as long as we care to look."

John Warner, *The Chicago Tribune*

OBJECTLESSONS

A book series about the hidden lives of ordinary things.

Series Editors:

Ian Bogost and Christopher Schaberg

Advisory Board:

Sara Ahmed, Jane Bennett, Jeffrey Jerome Cohen, Johanna Drucker, Raiford Guins, Graham Harman, renée hoogland, Pam Houston, Eileen Joy, Douglas Kahn, Daniel Miller, Esther Milne, Timothy Morton, Kathleen Stewart, Nigel Thrift, Rob Walker, Michele White

In association with

BOOKS IN THE SERIES

spacecraft

TIMOTHY MORTON

BLOOMSBURY ACADEMIC
NEW YORK • LONDON • OXFORD • NEW DELHI • SYDNEY

BLOOMSBURY ACADEMIC
Bloomsbury Publishing Inc
1385 Broadway, New York, NY 10018, USA
50 Bedford Square, London, WC1B 3DP, UK
29 Earlsfort Terrace, Dublin 2, Ireland

First published in the United States of America 2022

Cover design: Alice Marwick

For legal purposes the Acknowledgments on p. xii constitute an extension of this copyright page.

Library of Congress Cataloging-in-Publication Data
Names: Morton, Timothy, 1968- author.
Title: Spacecraft / Timothy Morton.
Description: New York : Bloomsbury Academic, 2021. | Series: Object lessons | Includes bibliographical references and index. | Summary: "Science fiction is filled with spacecraft. And in the real world, eager industrialists race to develop new vehicles to travel beyond Earth's atmosphere. Space travel can seem like a waste of resources or like human destiny. But what are spacecraft, and just what can they teach us about imagination, ecology, democracy, and the nature of objects? Furthermore, why do certain spacecraft stand out in popular culture? If ever there were a spacecraft that could be detached from its context, sold as toys, modeled, turned into Disney rides, parodied, and flit around in everyone's head-the Millennium Falcon would be it. Based primarily around this infamous Star Wars vehicle, Spacecraft takes readers on an intergalactic journey through science fiction and speculative philosophy, and revealing real-world political and ecological lessons along the way. Philosopher Timothy Morton shows how the Millennium Falcon is a spacecraft par excellence, offering readers not just flights of fancy, but new ground to stand on"— Provided by publisher.
Identifiers: LCCN 2021013852 (print) | LCCN 2021013853 (ebook) | ISBN 9781501375804 (paperback) | ISBN 9781501375811 (epub) | ISBN 9781501375828 (pdf) | ISBN 9781501375835
Subjects: LCSH: Space vehicles--Philosophy. | Ontology. | Object (Philosophy)
Classification: LCC TL795 .M67 (print) | LCC TL795 (ebook) | DDC 629.4701—dc23
LC record available at https://lccn.loc.gov/2021013852
LC ebook record available at https://lccn.loc.gov/2021013853

ISBN: PB: 978-1-5013-7580-4
ePDF: 978-1-5013-7582-8
eBook: 978-1-5013-7581-1

Series: Object Lessons

Typeset by Deanta Global Publishing Services, Chennai, India
Printed and bound in the United States of America

To find out more about our authors and books visit
www.bloomsbury.com and sign up for our newsletters.

For Claire and Simon

Turning and turning in the widening gyre
The falcon cannot hear the falconer

—W. B. YEATS

CONTENTS

ACKNOWLEDGMENTS

Thanks so much to Ian Bogost and Christopher Schaberg. Thank you Bloomsbury. Thank you Haaris Naqvi. Thank you Alice. Thank you Antonina Szram. Thank you Taylin Nelson. Thank you Nicholas Royle. Thank you Maya Kóvskaya. Thank you Harriet Harriss. Thank you Philippe Parreno. Thank you Laurie Anderson. Thank you Elizabeth Freeman. Thanks to the Coronavirus for disrupting life so much I finally got back into the swing of writing weird stuff. Also thank you to the virus for not killing me when I got it, that was polite. Thanks to Ben Rivers and to Gareth Evans for being amazing pals, and similarly Kathelin Gray and Leslie Roberts. You've all kept me on the level during the most difficult two years of my life. Thanks Henry Warwick, fellow Yes fan and underminer of the labor theory of property. My life's work is committed to blurring the boundaries between active and passive, medieval Neoplatonic Christian constructs that inhibit genuinely revolutionary thought and action. Thank you to everyone who this last year or so has helped me to know that much more deeply than I ever thought possible—the discovery that I am non-binary gender.

INTRODUCTION

SHIPS AND CRAFT

When I was eleven years old, I used to walk to school. It was about a mile and a half from Hammersmith Station, through the Tubeway made famous by Gary Numan's Tubeway Army (Gary Numan of "Cars" fame). My eleven-year-old boy's head found comfort in all sorts of obsessive behaviors as I traveled through an alien, alienating land of flyovers and concrete and loud train sounds, toward the other alienating land of a very posh private school. I was growing up on bare floorboards, the dusty, broken kind, not the cool chic kind. But my schoolmates had chauffeurs, and in one instance, a helicopter.

One of my obsessive comforts was making up spacecraft. I would inhabit them in the cockpit of my mind, describing their specifications to myself, under my breath. When someone found out, I was unmercifully bullied for doing so, but it didn't stop me. Spacecraft were safe. They could protect me and whisk me away from the alienation I was feeling. At home, I was obsessively reading my collector's edition magazines: there was one about *Star Wars*, which I had seen on its first release on December 27, 1977, at the Dominion Theater in Leicester Square, in London; and there was a magazine about *Close Encounters of the Third Kind*, which I had also seen very soon after its release. I can't count the number of models of spacecraft, real and fictional, that I owned. And books about UFOs. I still have those collector's editions. I'm looking at them now. Unlike a lot of things in my life from that time, they were and are almost pristine.

I was using spacecraft as protection and as escape vehicles in my head. And so, to this day, I've wanted to explore what that was all about. In particular that's because I'm sure I'm not the only one.

This is a book about the space vessels of our imaginations. This doesn't mean that I'm not interested in Apollo or the Space Shuttle or Soyuz or Sputnik. Far from it. The whole point about those vehicles is, they also originated in some dream of a human being. The Space Shuttle bears a vivid resemblance to the shuttle as imagined by Stanley Kubrick in the film *2001: A Space Odyssey* (1968). And to design real spacecraft, you need a big imagination. In 2016 I was lucky enough to join Pharrell Williams in New York City, in a conversation with NASA about what messages the next Voyager spacecraft would convey to extraterrestrial lifeforms. It always struck me that writing is like sending a message in a bottle into space, and that this is why writing is also a kind of listening, like the giant radio telescopes that listen for signals of alien life. Voyagers 1 and 2 are ears just as much as they are probes.

And that's not the only reason to write about these fictional space vehicles. Imaginary "objects" are also objects. Objects don't have to be palpable and capable of being burned or shot around a particle accelerator to be real. The vessels of our imagination are real insofar as we can all think about them, as differently as we do. We can buy toys of them. We can watch movies containing them. They can hop from one mind to another.

This isn't such a strange thing at all. In fact, it would be very weird if thoughts were only "symptoms" of the mind that had them. How on earth could we communicate at all? Consider an idea. It's not a naked thing. It always comes clothed in some medium or other. An idea is very like a tweet or a meme. It has a certain flavor, a certain size and shape, a certain speed and intensity, and a certain kind of format. Ideas don't just float around in the void. Nothing floats around in the void. Spacecraft of the imagination are just as real as Soviet spacecraft.

This is true for many things, but especially true for spacecraft. They are never simply "in" the mind (is anything ever "in" a mind?). Spacecraft are LEGO models, theme park simulators, replicas, things we make out of unembellished bricks and sticks at the age of seven.

Even pictures "in" the mind are independent of those minds, which is why I can say *Millennium Falcon* and you can know what I mean. At the beginning of the twentieth century, the philosopher Edmund Husserl had a striking idea. He was trying to prove that there was something fishy about nineteenth-century theories of logic. There was a theory called psychologism. It proved, so it thought, that logical thoughts, propositions or what have you, are true because they are symptoms of healthy brains. It was a very popular idea.

And what, I hear you ask, is a healthy brain? Well, it's a thing you can determine using science. And what is science? Well, it's a series of statements about patterns in data that can be shown to be true using logic. And what is logic? Why, it's a symptom of a healthy brain. And what is a healthy brain? Well, it's a thing you can determine using science. And what is science? Let me explain—it's a series of statements about patterns in data that can be shown to be true using logic. And what, pray, is logic? Why . . .

Houston, we have a problem.

Husserl concluded that logical propositions are independent of the mind.[1] They are their own things, like spoons or spacecraft or slugs. They are "objects"! (I'm putting *object* in quotation marks because I don't want to get started with all kinds of fancy notions about objects versus subjects.) It doesn't matter who is thinking them. You can be deranged and retweet something quite sensible. The tweet won't be affected by your derangement.

Then Husserl went on to say something truly amazing. He decided that if logical propositions could have an independent, object-like format, so could other "mental" phenomena such as hoping, loving, wishing, fantasizing, promising. Thus was born the philosophy of *phenomenology*. The basic idea is that things—ideas and wishes but also spoons and daffodils—come with ways of "having" them or accessing them, like memes or tweets. Nothing is naked in the void. So, things such as spacecraft are as real as things such as baseballs or cricket bats. Whether they're images in our mind, or models made of LEGO, spacecraft are independent of the minds or brains that are imagining them, or the LEGO bricks that are modeling them (and so on). If I say the word "spacecraft," you will visualize something that won't just be a symptom of your particular mind.

I love the word *phenomenology*. When in the mid-1980s I arrived in Oxford for my undergraduate degree, I had just discovered it and I said it a lot, quite possibly under my breath in a highly eccentric way, and people noticed and made fun of me for it. Phenomenology had a rather hard time of it for a few decades, but it's now enjoying a big comeback. I think it's because of two factors: a misinterpretation of Jacques Derrida, who was himself a devoted phenomenologist; and because of a resurgence of scientistic reductionism.[2]

Derrida developed deconstruction after a careful critique of Heidegger (who coined the term). Heidegger had himself launched a thorough critique of Husserl. What people tend to forget when they think about this is that both Derrida and Heidegger are doing "critique" in the way Kant means it. *Critique*, as in *Critique of Judgment*, doesn't mean "snarky criticism of." *Critique* means "loving, deep assessment of." Derrida is critiquing Heidegger because he *likes* his work. Heidegger is critiquing Husserl for the same reason. Derrida's critique of Heidegger is Heideggerian, wandering around inside a topic and eating it out from the inside. Heidegger's approach in his *Being and Time* is a phenomenological one, exploring the fact that *how* something arises tells you about *what* it is. Phenomenology isn't *wrong*— in a way, it's *more right* than we might suppose. "Deconstruction" is not destruction. It is, to use the coinage of one translator of Heidegger, a *destructuring*.[3]

Then there's scientistic reductionism. *Scientism* is a religion; despite the first two syllables, it's almost the opposite of science. Scientism forgets that a scientific fact is an interpretation of data, and supposes that this fact points to a reality that is *more real* than others. If science says we are made of atoms, scientism says that atoms are more real than human beings and rabbits. That's not true and it's not scientific. Scientists aren't really allowed to say what's more real than what. From their point of view, that would be like living in the middle ages, a time of religious authoritarianism.

Scientism reduces "phenomenology" to just one kind of phenomenology—a fancy word for "subjective experience of . . ." And we all know what *subjective* and *experience* mean (I'm being sarcastic). They mean superficial, unscientific impressions. The trouble with scientism is that some science is afflicted with it: it worsens the closer one gets to looking at human beings in a scientific way, in fact. Psychology is much more reductionist than physics—the really hardcore materialists these days are psychologists. It's psychology that has turned "phenomenology" into "subjective experience of . . ." Psychology gives phenomenology an even worse reputation than Derrida. We've gone from "the study of how things arise as a clue to what they are" (which is pretty much science) to "a series of superficial subjective impressions." There's no effort to explain what is meant by "subjective" or "impression" and no idea that "subjective" implies a whole lot of medieval thinking that psychologists should be embarrassed to have chittering around in their heads.

This book uses the insight that "ideas" and "images" are objects in their own right. Spacecraft aren't just symptoms of our brains, figments of our imaginations. They are autonomous beings. They have something to tell us. I'm also going to argue that the very concept of phenomenology has ever so much in common with the idea of *spacetime*, which Einstein discovered roughly around the same time, the very early twentieth century. One of the main goals of *Spacecraft* is to allow access to a felt experience of spacetime—not the empty void that we still too often assume space to be, but a liquid, luminous swirl.

The medium spacecraft travel within, the scintillating ocean we call *hyperspace*, is good old outer space, but imagined as a substance, a thing, an object! It's a peculiar and dangerous habit of western thought to picture a solid when we hear that word, "object."

For me, objects are all kinds of liquid. If hyperspace has any cultural roots, it's in African philosophy and spirituality, as the liquid that acts as a door or portal: the Kalunga of the Kongo Cosmogram, the ocean between the worlds.[4] I don't know whether or not the concept was directly appropriated by the major hyperspace shows (*Star Wars*, *2001*, *Doctor Who*), but the similarities are quite extraordinary.

Spacecraft and hyperspace go together like the proverbial horse and carriage. Spacecraft are the things that "get us" there. If things all have a form— if ideas are meme-like or tweet-like beings that have size, shape, vector, point of view, tone, color. . . then why shouldn't space itself be like this? This is a very politically progressive image, as I will explain, as well as a very cool thing to think about in itself. Hyperspace is utopian in both senses—we can't reach it technically yet, because we can't travel faster than light, so it's a *no-place* (Greek *ou* meaning "non"); and it's a door between worlds, not exactly a "place" as western thought imagines such things—Kalunga is KiKongo for *threshold between worlds*. But it's also a *good place* (Greek *eu* meaning "good"), a quiet, delicious, sparkling cream. I fell in love as soon as I saw the TARDIS rushing through the spacetime tunnel of shiny silver in *Doctor Who*, like the nacre of a gigantically elongated shell in the shape of a tubeworm.[5] (The cosmic portal of the Kalunga is imagined as the inside of a spiral shell.) Then I saw *Star Wars: A New Hope* (1977) at the age of nine, which confirmed and amplified the love.

In 1977 I was obsessed with aliens. Steven Spielberg's *Close Encounters* and eventually the Royal Institution lectures of Carl Sagan (and the subsequent book and TV series *Cosmos*) convinced me that aliens existed.[6] I still see 1977 weirdly as the future. Jimmy Carter had put solar panels on the roof of the White House. Andy Warhol's "worm" NASA design from that time kicks the ass of the "hamburger" of the Reagan years and later. Like hyperspace the "worm" is a smooth continuum, like a hip hop tag, which you are supposed to make without stopping the spray of paint from the spray can. Where did 1977 go, and is there any way to get it back? It's still the future. The UK was interesting then, too. Labour was still in charge. Punk and Pink Floyd's *Animals* (one of my favorite albums) dominated the airwaves.[7]

This book is a homage to a way of thinking and imagining that has nothing to do with mechanical nuts and bolts and know-it-all performances of masculinity. It has everything to do with spiritual, sensual liquids. That's what spacecraft are really all about. This is a feminist book about spacecraft and hyperspace. One of my favorite philosophers, Luce

Irigaray, has written on the connection between feminism and mysticism.[8] That's why a strong impulse behind this book is feminist. There is a long and rich legacy of feminist science fiction.[9] Moreover, there is one dominant way to understand the aesthetics of hyperspace: a feminist one.

Irigaray is central to the way I like to think about the things that the philosophical school of object-oriented ontology (OOO) calls "objects": humans, hedgehogs, and in particular in this book, hyperspace. I'm going to be discussing OOO throughout this book, so here I will just briefly mention a couple of things. First, ontology doesn't tell you what exists: we can leave that to science and history and so on. Ontology tells you *how* things exist. If a thing exists, how does it exist? That's the question ontology is answering. OOO tells you that if a thing exists, it exists in such a way that nothing, not even that thing itself, can fully grasp, access, affect or otherwise influence it. Nothing.

Think about it. If I bite a banana, I have a banana bite. If I lick a banana, I have a banana lick. If I peel a banana, I have a banana peel. If I write a poem about a banana, I have a banana poem. If I use a banana in an example of how OOO works, I have a banana example. If for some reason the banana develops the ability to speak and goes on a chat show, what it will say about itself won't be the banana either—it will be a banana interview. Or say the banana ends up in therapy: "I first realized I was a banana when I was being used as an example of OOO in this book about spacecraft by this crazy philosopher guy. It was very traumatic." That's banana autobiography. Even the banana can't access the banana banana!

Irigaray's idea of the non-patriarchal body as a multiple entity that doesn't have any way of being put into a monist or dualist box, doesn't have an inside distinct from an outside, has multiple entry and exit ports, cannot be grasped by patriarchal philosophy . . . all these concepts are directly what hyperspace is all about, I argue. The chapter "The 'Mechanics' of Fluids" in her book *This Sex Which Is Not One*, cruelly and unfairly mocked by Alain Sokal, is in fact highly congruent with how to think hyperspace.[10]

One of the first titles suggested for this book was *Spaceship*. I argued early on that we shouldn't call it *Spaceship*—we should call it *Spacecraft*. A spaceship is large. A spacecraft is small—if it's a boat it's not an Atlantic steamship or an aircraft carrier. A spaceship has a consistent crew. In *Star Wars*, the Empire's vessels are quite obviously ships. They are part of an official fleet. There is a large and meticulous class hierarchy on a ship. There's a captain and a crew— if you're on a ship, the captain is at least the master, or lord, lady, duke, duchess, of the ship, in charge of punishment and marriage and burial. In space, you see Captains Picard and Kirk officiating at weddings. You hear a whistle that announces that the captain is on deck. A spacecraft is some kind of speedy yacht or catamaran. The work in a spacecraft isn't a job. It's a passion.

You just climb in and fly the *Millennium Falcon*, preferably as fast as possible. We can say the same of the *Heart of Gold*, the ridiculously fast spacecraft that is at the center of Douglas Adams's *The Hitchhiker's Guide to the Galaxy*. There's that sense of "craft" as skill, in the sense of a technique that you learn, and in the sense of being "crafty" like Odysseus—his stories are a little bit like those of the pilots of the *Falcon*. Scylla and Charybdis features in the prequel *Solo: A Star Wars Story* (2018) in the form of a space monster and a black hole.

Consider a scene in *Close Encounters*, which was released shortly after the first *Star Wars* film. Roy Neary is sitting at a train junction. A car seems to pull up behind him. It speeds off and the driver swears at Roy, who is stationary, waiting. Then another seeming car pulls up. The lights rise slowly above Roy's truck. It's an alien spacecraft. Then the Mothership arrives— now *that* is a ship. It's not a "mother-craft." Roy's truck is a craft, and so are the cars and spacecraft that pass him, horizontally and vertically. The train somewhere on those tracks and the Mothership aren't craft. They are ships. Consider the seriousness with which Elon Musk has named his vessel the Starship. It's quite serious. Then contrast that with David Bowie's rickety spacecraft in *The Man Who Fell to Earth*.[11] They're quite different.

The difference between ships and craft is about parts and wholes. I am very interested in holism, because I'm an ecologist. You need to believe in wholes to be one, wholes such as meadows, oceans, habitats and cities, ships and craft. But what kind of wholes are they? There is a *ship set theory* and a *craft set theory*. Ships imply a kind of holism where the whole swallows all the parts like Pac-Man or like how water dissolves salt crystals to make salt solution. This is surely why we talk about the ship of state. We don't talk about the craft of state.

Craft set theory is a whole new kind of holism. Crafts contain things but not entirely. You can hop on and off. You're not a uniform-wearing crew member. In a craft, and I know this is going to sound weird, the whole is always *less* than the sum of its parts. We usually think that wholes are always greater. But this is just ideology. I believe that for something to exist, it means that it exists in the exact same way as something else. This is the basic tenet of OOO, which I've already started laying out by talking about how ideas and images are things in their own right. If there are football teams, then they exist in the exact same way as football players. One is not "greater" than the other. There is one team. There are eleven players. Therefore *the whole is literally less than the sum of its parts!*[12] This is such a simple concept. Why does it sound so weird? It's because we're indoctrinated by an idea that wholes are greater than the sum of their parts. But that leads in the end to fascism—where one's identity is totally defined by the bunch, bundle (Latin, *fasces*), gang, squad into which one is incorporated.

We often assume that cars and craft are expressions of individualism. That indicates a failure of imagination. We can't imagine wholes as anything other than these fascist "solutions" (in every sinister sense). Cars are seen as "expressions of personal freedom" as the lead character says in David Lynch's *Wild at Heart*, a phrase imitated by one of UK Prime Minister Margaret Thatcher's ministers in defending cars against trains.[13] But now we can imagine craft as *wholes*, as expressions of holism. Part of the fun of this book is stealing the *Falcon* back from the Reaganite idea of individualism versus the "evil empire" of Soviet Communism. Alluding to *Star Wars*, President Ronald Reagan called the USSR the "evil empire."[14]

The *Falcon* doesn't fit into neat, clean, official forms of commodity circulation. It's a craft whose lot in life is to be stolen, borrowed, rented, won, lost, discarded . . . You don't buy it or sell it. You chance upon it. Then you get inside it, learn how to cooperate with it, and escape somewhere. Doesn't this mean that the *Falcon* departs from the individualism that bankrolls the form of capitalism during which *Star Wars* was created? Doesn't that mean that the *Falcon* might be from the future, indicating ways of organizing how we enjoy stuff that are different from neoliberal capitalism?

"How we enjoy stuff" is a pretty good translation of the word "economics." If I give you two dollars, you can buy this pint of milk (it's an expensive supermarket, I'm afraid). Economics isn't really about dollars and business cycles. That's the "alienated" version of it. It's really the "house customs" (Greek, *oikos*, house; Greek, *nomos*, customs). Economics depends on ecology. Ecology is the logic of the house (Greek, *oikos*, house; Greek, *logos*, logic). The house could be the biosphere, your neighborhood, a habitat . . . or all three of those, overlapping. If you have ten bucks, you can fill up your tank with this much gasoline. The gasoline involves extracting stuff and hurting lifeforms and building refineries next to where nonwhite people live. It's all about who gets to enjoy (and who gets not to enjoy) the world. The reason why you can exploit people and drill for oil is that there were fossils that got compressed into oil in the substrata of Earth's crust, because evolution . . . and so on.

We always necessarily find the *Falcon* somewhere. It's never just floating about in a void. It's always located. It implies a world—an economics and an ecology. Someone won it, thinks they own it, parked it, put clamps on it, some gigantic worm swallowed it . . .

Ships are *docked*, in convenient places that enable them to leave efficiently—they are part of very specific and unified ways of organizing things: "Rule Britannia, Britannia rule the waves . . ."[15] It's as if ships are part of a world so integrated that it has disappeared. It's everywhere. It's the evil empire.

Crafts however are *parked*. You have to find a place to perch. Crafts are symptoms of wholes that are ragged, fuzzy, vague, porous. You can come and go. You can slip out from underneath.

If we want to think about global economics and ecology, we need to think in wholes. We need to think the opposite of Prime Minister Thatcher who said, "There is no such thing as society."[16] But we need to think of wholes as "less," as fragile and contingent. Ecological beings such as lifeforms and ecosystems and the biosphere are also wholes whose parts outnumber them and are disparate, not "greater."

Humans are made of all kinds of nonhuman bits and pieces—it's called evolution. Meadows are also made of all kinds of bits and pieces that they share with other ecosystems next door to them, such as clouds, streams and small vole-like creatures scuttling about. In a way then, ecological beings are crafts, rather than ships.

Ecological beings are *Millennium Falcon*s.

The *Falcon* provides a way to think about ecological science, politics, ethics and art . . . and pleasure. Ecological thinking is usually stuck in shock mode, and usually comes in the form of oppressive information dumps and jeremiads about how evil you are. Wouldn't it be more fun to think about Chewbacca and Han Solo and Leia and Rey and Lando and Luke and so on? This book is predominantly about the elephant in the room: the most popular, corniest, well-known series of movies.

There is a naïve, sincere way of talking that is a part of OOO talk. It's quite like *Star Wars* or *Sesame Street* or *The Muppet Show*, which are all related, to the extent that *Star Wars* has always been *Muppet Star Wars*. Actual Muppets from Jim Henson's studio, actual Muppet voices (such as Frank Oz who plays Yoda) demonstrate this link (first in *The Empire Strikes Back*, 1980). But there's something deeper: both projects are about using seemingly naïve, simple, sincere things to make profound social and philosophical points. Both are about democracy, the idea that anyone can have access to these points, that anyone can make them. They are also about expanding democracy to include more than humans.

The Muppet parodies of space movies and television feature ships rather than craft. *Muppets from Space* is the movie, and "Pigs in Space" was *The Muppet Show* series.[17] Once they show up in *Star Wars*, they often pilot craft. Yoda's craft is a good example of what I am going to be calling a *coracle*, a specialized survival or pilgrimage craft for one. Consider a *Falcon*-related scene: Greedo talking to Han Solo in the cantina. There is no explanation, no mediation. You are just there, overhearing a conversation between two lifeforms, and it's clear that Han could understand Greedo's language, a sign of some kind of cultural understanding and acceptance and a sign that there isn't a universal language that everyone has to speak. Calling Greedo or any other being in *Star Wars* an alien would be a big mistake; so would be calling a Muppet a puppet. Sure, technically they are operated by humans. But they have such a life of their own. The point of looking at the Muppets is not to gasp at the skills of a puppeteer. It's to find out what Animal is going to do in *Muppets from Space*.[18] There are no puppets or aliens in *Star Wars* and *The Muppet Show*. Neither are about human mastery. In lots of ways, they are both about how the human idea that humans are the masters is mistaken and dangerous.

For quite a while the dominant way for people like me to be "right" has been called *cynical reason*.[19] One sees through the other person's point of view rather than exploring it. It has a bad side-effect. For example, I can call you out for not being as left-wing as me, because you don't realize just how all-pervasive ideology is. If I can convince you how paralyzed you really are, I must be very intelligent. It's like a game of Pac-Man where my job is to munch down as many other arguments as possible. This can be a disaster if both of us are trying to change the world for the better. Since when did disempowering people become a great tactic? Writing in a deliberately naïve way is a very interesting experiment in countering this.

The other side-effect is that one's argument sounds like it is coming from outer space. I am sitting in judgment on how other arguments are hypocritical, how hopelessly compromised and flawed they are. But there I am, saying that—while my own argument is flawed and compromised because I'm pretending to be on the outside. If words, thoughts, and ideas come in the form of pixels, tweets, memes, pencil drawings, jottings on a notepad or newspaper articles (and so on), there is no way to have a totally naked idea that isn't physically embodied in some sense.

Well, let's take this idea out for a spin. Let's think about those spacecraft that show up in our heads and in magazines and books and movies. I reckon there must be several spaceships and spacecraft in the human cultural imaginary. This isn't an exhaustive list at all:

The ark

The juggernaut

The frigate

The fighter

The explorer

The yacht

The *machina cum dea*

The coracle

Yet none of them are exactly the *Millennium Falcon*. That is why I'm going to make that spacecraft the centerpiece of this book. The *Falcon* can become any of these types of vessel—it can even be a pirate-ship version of the frigate, or a miniature juggernaut, ploughing through towns and rock faces. And each of these vessels can become other ones. But none of them can be the *Falcon*. That's because the *Falcon* stands for the irreducible uniqueness of how things are.

Let's consider these vessels one by one:

The ark. The strongest examples would be the ark-ship carrying all remaining life forms in *Silent Running*; the bone-shaped Jupiter ship in *2001*; and the *Endurance* containing human embryos in *Interstellar* (2014). There is also a powerful episode of *Doctor Who* entitled *The Ark in Space*.[20] The *Falcon* can become an ark for surviving members of the Resistance.

The juggernaut. The destructive chariot of Hindu mythology provides the name for this vessel. The Death Star and the Imperial Cruisers are juggernauts, as is the militarized version of the Starship Enterprise in the second reboot, *Into Darkness*. As it shaves bits of building and rock, the *Falcon* is a miniature juggernaut.

The frigate. Frigates are warships, such as the Starship *Enterprise*. To serve its initial purpose as a suspect trading vessel, the *Falcon* is a frigate, equipped with weapons to prevent its cargo from being stolen or damaged.

The fighter. The small craft that Obi Wan flies in the *Star Wars* prequels comes to mind, along with the X-Wings and Tie Fighters of the earlier and later films. They're only big enough for one or two passengers.

The explorer. They are often shaped like wheels, such as the space station that orbits the planet *Solaris* in Tarkovsky's film (1972). In the film *Interstellar*, the *Endurance* and its companion craft, the Rangers and Landers, comprise an explorer (versions of which are being built for real by NASA). Or consider the *Event Horizon*, whose gyroscope-like hyperdrive folds space and unfortunately summons hell beings into the bargain.[21] The wheel-like space station in *2001* counts as an explorer, insofar as it has been built to house people who are going to visit the monolith on the Moon.

The yacht is a lower-key explorer, a more mundane one such as a touring vessel, like the space-going Titanic in an episode of *Doctor Who* featuring Kylie Minogue.[22]

The machina cum dea. I'm inverting the phrase *deus ex machina*, "god from the machine," a term from Greek drama. Consider the motherships of UFO fame, for example the ship full of Gonzo-like beings in *Muppets from Space* or the aliens who temporarily whisk away the crucified Brian in *Monty Python's Life of Brian*.[23] Aristotle warned against employing a *deus ex machina* in a drama—which is why the alien rescue of Brian turns out to be an illusion.

In Aristotle's day, the *machina* was a platform on wheels on which would stand a god or goddess doling out judgments, like a cosmic referee. Aristotle thought that they violated the "unities"—the fact that a play that is more integrated in terms of time, place and "action" is more intense. This is true, I think, because at its most basic, a play is a dance: dances are in the moment. Unity of time means you never leave the present moment; there are no cuts or jumps. Unity of place means you never leave the moment in another way—there are no scene changes. Unity of action is like how dancers can only do what human bodies can do. Jealous characters should behave jealously and mountains and trees should not violate the laws of physics.

The Doctor's TARDIS is everywhere all at once, violating unity of time. It is notoriously "bigger on the inside"—infinite in fact—violating unity of place (then there's all the different worlds the Doctor visits in each episode). The TARDIS looks on the outside like a police telephone call box from the 1960s, because its "chameleon circuit" cloaking device got stuck. The TARDIS is a living breathing violation of unity of time and place. Surely the Doctor is a *dea ex machina*? No. As the lead character, she isn't the headmistress who grades everyone at the end. The Doctor always gets caught in the action. Since the Doctor is just a regular person, albeit an alien of great intelligence, she gets angry, confused, makes mistakes, gets too involved or too little involved. Likewise, the *Falcon* often functions as something like a *dea ex machina*. It swoops in to the rescue at a crucial moment. Only it doesn't seem to distribute justice in an absolute way. It seems, rather, to open up the space for things to happen. Its version of justice doesn't come from outside the universe, but from within it.

The coracle. Some spacecraft are *spiritual.* The craft could resemble the *coracle* of old, in which hermits sailed to holy lands. A coracle is nowhere near big enough for a hierarchy, which literally means "rule of the priests." Coracles are vessels of mystical experience, the kind that doesn't depend on priestly mediation. One appears toward the end of Samuel Taylor Coleridge's *The Rime of the Ancient Mariner* (1797). It rescues the Mariner from his "ship," on its last legs from its journey through a netherworld. And what a netherworld—it sometimes feels very like the rich disturbing beautiful flickering of hyperspace:

> And soon I heard a roaring wind:
> It did not come anear;
> But with its sound it shook the sails,
> That were so thin and sere.
>
> The upper air burst into life!
> And a hundred fire-flags sheen,

To and fro they were hurried about!
And to and fro, and in and out,
The wan stars danced between.[24]

You could only really return from such a disturbingly magical place via a coracle. A ship is way too clumsy and liable to invasion and corruption. Likewise, the EVA pod of *2001*, the little spherical vessel in which Dave is hurtled through hyperspace is very much a coracle. Coracles are famously round. Coracles voyage through spiritual liquid. The small sphere in which Jodi Foster passes through a wormhole in the movie version of Carl Sagan's *Contact* would be a good example of a coracle, along with the spherical EVA that Dave flies into the monolith's hyperspace in *2001*.[25]

As the craft carrying Carl Sagan's curated humanity data into the cosmos, The Voyager 1 space probe is a coracle. What Sagan was able to with its photograph of Earth, the "pale blue dot," speaks to this.[26] Just as it left the solar system Sagan had the probe turn around and take the last possible picture of Earth—the last because it's just one pixel large. Then he improvises extempore on the tininess and pettiness of human aggression amidst the cosmic wonder, what Percy Shelley aptly called "a wilderness of harmony."[27] Voyager I becomes a spacecraft in *Star Trek: The*

Motion Picture. A gigantic one, a coracle gone insane, blown up to contain virtual visions of everything it has encountered. "V-Ger" has surrounded itself with a strange spiritual realm, all its data turned into a curvilinear cloud with no exit that looks very like the hyperspace on the cover of this book. It wants to know what it is. It is on a deeply spiritual journey.

As I said above, that unique spacecraft, the *Millennium Falcon*, can become all of these vessels. I have organized this book in terms of how we encounter it. So it's logical that our first chapter should deal with the fact that it's mostly a "hunk of junk" as Lando Calrissian puts it—or, more succinctly, garbage.

1 GARBAGE

"That one's garbage," shouts Rey. She's running with Finn, an imperial Stormtrooper with PTSD, toward a craft, a Quad Jumper that will help them elude a swarm of attacking imperial Tie fighters. Right as she says it, the fighters blow up the Quad Jumper (*Star Wars: The Force Awakens*, 2015). "The garbage will do," yells Rey, and the camera turns to the garbage— lo and behold, it's the *Millennium Falcon*.

The *Falcon* is the ultimate found object in the world of *Star Wars*. Indeed, the *Falcon* stands for the *found-ness of all objects whatsoever*, the fact that they resist total appropriation.

If you could salvage just one thing from the *Star Wars* franchise, what would it be? I vote for the *Millennium Falcon*. That seems to be how the makers of the reboots feel about it—let's not forget they were part of the audience at one point. The most potent form of fandom would be—in fact, is—actually creating another episode of the story. Drama has always been about the Chorus. We think of the Chorus only as a group of commentators. But the Chorus are really the audience, reflected onstage. In the rave-like rituals from which drama arose, that would be literally true: we would all be onstage. Modern drama and cinema with their curtains and screens often feel like an objectified "thing" we are merely witnessing. But *Star Wars* wouldn't exist unless Lucas thought that you might want to see something like it. To that extent, the director is also a member of the audience, just like how I feel like the first reader of these sentences while I'm writing.

The director and the writer are like pilots of the *Falcon*, who always have to figure out how to do it on the fly. Free will is overrated. We think that *active* means *definitely not passive*. But this is just a patriarchal tweet from medieval Christianity, which parceled up the world into souls versus bodies, active versus passive. Because the *Falcon* "tells you" how to fly it, Han Solo doesn't have a monopoly on how to fly the thing. We (including the makers of the franchise) gradually learn that he wasn't the first owner of the spacecraft. Because he isn't totally in charge of it, Rey is able to modify it: "I *bypassed* the compressor!" she tells him proudly. You have to read or direct the Falcon—you can't author it or master it.

Pilot in Greek is *kubernetes*—from which we get the words *cybernetics* and *governor*. But governing in this case doesn't mean you are the ruler. It means that you are going along with the sails and the ocean, steering and veering. When you are veering, are you doing it or is the ocean doing it to you?[1] The tractor beam moment in the first *Star Wars* film is a vivid example of what always happens anyway. We are always caught in a number of different tractor beams. Try to drive totally consciously or "actively." You will die in a horrible accident. You need to be in a mild state of hypnosis, as if you were being sucked forwards by the road and the other cars.

You can't be in charge of knowing what will come next. The future can't be totally predicted. Han Solo says something really, really remarkable and deep about this in episode 7. He's about to launch into hyperspace from inside a ship. Rey says "Is that possible?" and Han replies, "I never ask myself that question until after I've done it." After you've "done it" you might ask that question: *how on earth did I manage to do that?* It's as if the action is also a thing you can't quite grasp or know totally. You're sort of looking at it as if you were turning a present around in your hands, wondering what's inside the wrapping.

It's moving when Rey says "the garbage will do" and we realize, as the camera pans left, that she is talking about the *Falcon*. Rey and Finn and George Lucas and J. J. Abrams and we audience members all want to "pilot" the film. The garbage turns out to be a wonderful gift. The ultimate present would be something you already had but didn't realize, something that you had thrown away as garbage, but it was still there because you forgot to empty the bin.

This is also ecologically true. There is no "away" in the space of art, just like there's no "away" at planet scale. *Away* just means that your waste just became someone else's problem. The *Falcon* is a kind of shit that is actually gold. "Away" is just a word for the stuff that's to the right of the camera angle. It's not away at all, you just have to orient your face and . . . lo and behold, wonder of wonders, it's the *Millennium Falcon*. You go "Is that even possible?" and realize the truth of what Han says—before he says it.

Rey knows about craft. She doesn't have any money or family. She scavenges for a living. She's like a crow—or maybe she's like a quintessential human. Humans are scavengers, carrion animals. You can imagine humans feasting on bits and pieces left behind by lions. We've already seen Rey piloting something like the *Falcon*. It's when she sits on a thin rectangle of metal and slides down a gigantic sandy slope. So we know she's going to be adept at it. We already know that the *Falcon* is perhaps a surfboard, or maybe it's a frisbee.

There is something very significant about all the vehicles in *Star Wars*—all the good ones anyway. I noticed it at once when I saw the first film in 1977. *They are dirty.* This was certainly not a feature of the spacecraft I had seen previously, such as the *Starship Enterprise*, the pristine vessels of *2001* or the space ark of *Silent Running*. The space station in Tarkovsky's *Solaris* is also quite spick and span by comparison. What did these dirty spaceships mean? It meant they had been around, before I looked at them. They were marked with encounters of which I had no clue, the residue of other interactions. They were definitely physical, not just figments of my imagination.

We often imagine the future and "outer space" as places without dirt. Dirt is indeed a sign of the past—something's happened to your bath and you haven't wiped it off yet. So perhaps very simply we assume that the future must be the opposite—totally clean. We then add to this all kinds of prejudices about modern life. Before "now," things were a lot dirtier. So after now, they ought to be a lot cleaner. This is just building our expectations—which are themselves the past—out into a predictable possibility space that we call the future. It's not really the future, not in terms of something genuinely new, something that we can't predict.

The dirty spaceship isn't just an indicator that it came from a past we can't access, quite thrilling above and beyond the realist aesthetic, the "gritty" realism as we like to say. The dirty spaceship lets you know that there could be *an unpredictable future.* The dirt is utopian! It's not a mark of failure at all. Or rather, it shows you how failure can be a gateway to something different. The things you don't expect or predict, the contingencies of life—the "dirt" or "noise" in the cybernetic sense, the stuff that seems to get in the way—have a magical quality.

Think about listening to an LP record, if you can. You get to a track that you don't like, but the record is too far away to change and anyway, you might damage the stylus if you lift it up and plonk it down clumsily. So you listen to the tune. This isn't Spotify or iTunes where you can determine in advance what you're going to hear, or skip over what you don't like to get to the tune you do like. After repeated listening, the tune *grows on you.* Something new has happened. What you endured has turned out to be something rich and creative for you. The garbage you tried to ignore, on closer inspection, turns out to be a treasure trove.

Life and contingency are like that. You don't actively program everything in advance. You can't.

You can never tell in advance whether your program will really work or not, or whether it will succeed where others might fail. If you program too much, you shut down the possibility of improvising, which also has to do with being "passive" rather than being "active." Or perhaps it's better to say that the binary is suspicious. Acting is based on attuning or attending to a situation. Playing music with others is based on listening. Writing this book means me "listening" for what you might like to be reading right now. I'm reading this sentence, just as much or more than I'm typing it. Moreover, the concept of free will often sets up a dualism between the being with free will—who can do anything to anything—and the "anything" side of the equation. And fundamentally, this relationship is one between master and slave. The work of Denise Ferreira da Silva demonstrates this powerfully.[2] So it's about time we took a serious look at the active–passive binary.

An old but useful definition of dirt is "matter in the wrong place" (Mary Douglas).[3] That means that you programmed your world a certain way, and that the boundaries have to do with property, propriety, appropriation, the proper. But matter in the wrong place is just a track on a record that you endure and might end up growing on you. Matter in the wrong place is the dirty *Falcon* covered with a gigantic sheet that you think you should run past because it's "garbage." But it turns out the garbage will do.

What does *in the wrong place* really mean? Well, the matter must be in the wrong place *for* someone. I rent a house that was built in 1929. The kitchen floor is made of wooden slats, painted grey. It's very easy for dirt get in those slats, meaning stuff that I don't want in those slats. I like cleaning, laundry, tidying, doing the dishes. Most of my day is spent doing those things, and that's also how I have my ideas. So quite often you'll find me on my hands and knees cleaning in between those slats in the kitchen (and the laundry room) with a sponge.

Garbage is stuff that isn't *for* someone anymore. The dirty slat isn't how I want the kitchen. I want the kitchen to look good for me and for my human guests, and probably even my cat Oliver appreciates some clean kitchen slats near his bowl. Garbage therefore tells you something deep about what it means to be a thing. Oscar the Grouch of *Sesame Street*, or Diogenes the cynic in his barrel, or Hamm's parents in dustbins in Samuel Beckett's *Endgame*, are all symbolic of what the downed spacecraft also symbolizes. Objects don't just exist "for" some other being such as a human or a cat. They don't just exist in the eye of, or under the sponge of, the beholder. They do what OOO calls *withdraw*, which doesn't mean they shrink back in space, but that they exceed our ways of handling them—or any way of handling them.[4] Thinking about them, eating them, licking them, won't get at the thing in itself.

But that doesn't mean that there's some magical floor underneath all this that keeps going and going, like the cartoon character who stays alive no matter what you do to it. That's a sadistic and also a racist (and misogynist) idea of matter that is the default in white patriarchy. Think about it. To be a thing is to be *fragile*. Things can be destroyed. They can die. We care about the *Falcon* because it could fall apart—it certainly acts like it might. That's why we're happy when Rey and Finn make off in it. Its corrupt and seedy previous owner Unkar Plutt runs out, shakes his fists and yells "That's *mine*!" But it clearly isn't.

This default "substance ontology" implies that there is something "standing under" appearances (*sub-stance* contains the words for *under* and *stand*). The OOO vision is much more radical. What a thing is can be found nowhere else than in its appearances. They are inseparable from it. But the thing is not reducible to its appearances! A spacecraft is not a bucket. However, whenever we look for the spacecraft in itself, we find spacecraft data tailored to our specific being, our needs, goals, expectations, ideas of what "true" means and so on.

What to do with the stuff we call junk or garbage is exactly our ecological problem today. From plastics to carbon emissions, humans have accumulated a lot of garbage, a lot of matter in the wrong place. Knowing that things are not deeply defined by what they are "for" seems essential to forming a less violent, manipulative, and polluting relationship with the other beings on this planet. To go deeper with this thought, we need to consider the *Falcon* as pure contingency, as something that just happens to you, garbage or not. We need to think of its status as something you *win*.

2 WINNINGS

If you can lose a thing or throw it away, you can also win it or happen across it. This is another dynamic of the spacecraft in general, and of the *Millennium Falcon* in particular. It's also true of *The Heart of Gold*, the craft stolen by Zaphod Beeblebrox in *The Hitchhiker's Guide to the Galaxy*. Some lucky, charismatic person just comes upon this kind of spacecraft. The *Falcon's* original owner is the very pinnacle of charisma, and also appears to have won it, in a card game: Lando Calrissian.

The *Millennium Falcon* refuses to be a commodity with an objective price, something determined in advance relative to the other commodities that people use. We have seen how it is *garbage*, a thing that has fallen out of the commodity exchange system. Now let's dwell on the fact that it is also *winnings*. The *Falcon* is something you win in a card game, or not. You don't work to win a card game—you play the hand you are dealt, so there is a whole bunch of your supposed free will that the card game has suspended.

No matter how good you are at reading the player across from you—the real trick of playing a card game like poker—you are always at the mercy of the randomly-dealt cards. This can work against you but it can also work for you. Say you possess nothing in particular, but you happen to be a good card player, like the young Han Solo. You enter a seedy bar. You sit down opposite Lando Calrissian. You've seen him before. He cheated at cards that time. This time you have a good hand, and you win. You didn't have the wealth to buy the *Falcon*. But now you own it, because the *Falcon* was placed into a gift economy, rather than one of commodity exchange.

From the point of view of strict buying and selling and making a profit, there is something excessive and risky about the gift economy. Often they are indigenous. Gift economies rely on ceremonies such as the potlatch, a competitive giving-away of possessions. Whoever can give away the most is the winner.[1]

There are several examples of spacecraft that are won or stolen. The Doctor steals the TARDIS and flees from Gallifrey, his fraught home planet. The dissident Blake steals the *Liberator* in the brilliant Terry Nation series, *Blake's 7*.[2] Each spacecraft is a getaway vehicle—but in each case the real crime scene that is being fled is perpetrated by the state. The *Liberator* and the TARDIS are outfitted with very special powers. The TARDIS is technically everywhere, its interior infinite, and it's capable of traveling anywhere in space and time. The *Liberator* is equipped with a very powerful computer, Zen, whose giant octagonal screen pulses with orange light as the voice of a wise and pompous old man.

Spaceships are made by the state, and you have to earn the right to fly them. Spacecraft are stolen or won and you have to learn to fly them, on the fly. Because the state does not exhaust the things it controls, they can slip out from under its shadow. As a lowly, though beautiful, freighter, the *Falcon* is able to leave the remote, flea-bitten spaceport of Mos Eisley because the state—in this case the Galactic Empire— can't control everything.

Spacecraft as winnings demonstrate that the state is fundamentally weak. The state requires spaceships, laser canons and planet-destroying tech not because the state is all-powerful, but because it isn't. Spacecraft thus tell us something about how wholes are always less than the sum of their parts (see the Introduction).

What links the previous chapter and this one is *luck*. "I can do this. I can do this": starting up the *Falcon*, Rey is also trusting her luck with the audience that she can be a good, functioning character in *Star Wars*. Coming across garbage or winning something are all about contingency. Luck is either in the eye of the beholder, or it's not. In a truly mechanical, Newtonian universe, there is no such thing as luck. But in a universe where quantum theory explains practically everything, luck is baked into the structure of reality itself. *The Heart of Gold* runs on an "infinite improbability drive" that depends on generating the greatest amount of improbability possible and thus occupying every point in the universe at once, "without all that tedious mucking about in hyperspace."[3] It arrives at its destination by rapidly narrowing down the probabilities until, as Trillian puts it, "We have normality."

I sometimes think that the word *deserve* should be banned. When something nice happens to you, your friends are liable to go "Congrats! You deserve it!" Why they can't spell *congratulations* in an age of light-speed internet, I have no idea. I had always thought that the internet, being so fast, was a great excuse to make like a character in a Jane Austen novel and write really polite, formal messages to people. But people seem to have gotten caught in the speed. Speedy aggression aside, the trouble with "Congrats! You deserve it!" is that it's theistic. It assumes some kind of mechanical determinism operating behind the scenes. The problem is that, if you deserve the great thing, then you also deserve the terrible thing that happens tomorrow. For example, there is no particular reason why I rather than someone else is a little bit successful. Conversely, there's no particular reason why I am a survivor of various kinds of abuse. It just happened.

Let's consider the opposite of luck. In Western philosophy, this opposite was born right around the time of Newton, in the thinking of John Locke. For Locke, luck has nothing to do with owning a thing. You have to work on it to own it.

"That's *mine!*" yells Unkar Plutt, the supposed owner of the *Falcon*. He's been oppressing Rey forever, giving inadequate portions of nasty food in exchange for parts she plunders from a crashed Imperial Cruiser. He's yelling because Rey and Finn are flying away in something he considers his own. He shakes his fist impotently because working on something or appropriating something *doesn't* mean you fully own it. Someone can always steal it. The anarchist Pierre-Joseph Proudhon was right to observe that all property is theft.[4] America, the Moon, Australia, Ireland—Anglos in particular have had a knack of poaching large amounts of real estate. And as Proudhon observes, the very concept of a slave implies appropriation and murder.

Another theory of property has to do with labor. It comes from John Locke. If you work on something, it's yours. The trouble is, there's a limit to working on something. It's called destruction. You own something insofar as you can destroy it. Locke tries to dress the idea in a smart costume, but ultimately we land in the same place as Proudhon and Unkar Plutt. Treating things as property means you have a right to destroy them. The old patriarchal saying *I gave you life so I can take it away* is a very disturbing version of this: the father has a right to kill their own child because they "worked on" giving them life. Sadism underlies this theory of value. "Object" all too often means something a "subject" can destroy. The Marquis de Sade is all about who has the right to destroy whom, to reduce them to the status of a thing that can be destroyed.

But Unkar Plutt doesn't own the *Falcon* exactly like this. He just acquired it from someone, who acquired it from someone and so on. It can be stolen. So the *Falcon* is a symbol of something that shouldn't be destroyed just because you own it. Someone else could make off with it. It's not like an "object" in that sense. Unlike the downed Imperial Cruiser, it's not just stuff to be plundered, eventually resulting in no Cruiser at all.

Unfortunately, both capitalism and Marxism rely on the labor theory of value. This concept of property also affects the very idea of who we are as human beings. The idea of an object versus a subject is also the idea of a slave versus a master. As soon as you have that dualism, you have a slavery situation, as Da Silva has argued. The *Falcon* is saying something about things that is quite different than normal subject versus object, master versus slave dualism.

Throughout the *Star Wars* series, the *Falcon* persists, no matter who is flying it. The *Falcon* is very definitely not property. Perhaps the best you can say about it concerning yourself is that it's *home*, like when Han Solo and Chewbacca re-enter the *Falcon*, having scooped it up as Rey and Finn fly from the planet Jakku. "Chewie—we're home," says Han, poignantly. "Home" is a strongly ecological concept. It also underlies the idea that property is something you appropriate, by taking it or working on it in some more sophisticated way. You pick a piece of fruit because you live near the fruit tree. So really the idea of property depends upon an idea of living somewhere—and living always implies a "somewhere."

The *Falcon* seems to have an almost magical ability not to be destroyed. It seems to be made of substances that graze and break other things, but that don't break when they touch them. The hint that you can't destroy it even if you try is part of the scenes where Rey bounces it along and over rocks, buildings, other ships. And the whole idea that if you work on something, it's yours, is clearly troubled by the *Falcon*. No matter who flies it, they don't really own it. You can encounter it. You can win it. You can find yourself suddenly running into it to escape from Stormtroopers firing at you. when you do, it's not really yours. It's like one of those bikes you can borrow from the city council. The difference is there's no city council. There's no ultimate owner of the *Falcon*. So what is the *Falcon* telling us about how you can "own" things?

As garbage, the *Falcon* is stuff you don't (want to) think about. It's the not-(wanting to)-think-about-ness of stuff. Since its inception, civilization had garbage places, places where corpses were buried, toilets that took your waste to a magical place called *away*. Now we know, because of planet-scale ecological awareness that there's no such thing as away. Away is just another place that you're trying not to think about too hard. *Star Wars* gave us planet-scale awareness in the later 1970s.

This ecological lack of an *away* is about the shocking ways in which remainders and excessive parts of our world now come back up into the toilet bowl of our imagination to haunt us. But the idea of *away* points to a deeper kind of hiddenness. Wherever you go, there's always something around the corner. Then you get around the corner and there's something around *that* corner. The more you see of the whole planet, the more this deeper kind of *away* actually manifests! You can see maps and schematics and photographs from space and accounts of Space Shuttle pilots looking at the aurora encircling the Earth. All these things aren't the actual Earth as such. They're pieces of it, or reports about it, or photographs or maps of it. The more you know, the more mysterious a thing becomes.

This is another example of what OOO calls *withdrawal*. You can't see where the manifest aspect of a thing stops and the withdrawal aspect begins. It's like looking at a Moebius strip. Where is the twist in the strip that makes your walking fingers suddenly flip over onto the "other" side? The correct answer is that *the twist is everywhere*. There is no "other" side—like there's no "away" at planet scale—because the twist in the strip occupies the entire strip. You can see the whole thing, you can hold it in your hand, and that's why it's so weird. It's not like there is a hidden trick. The trick is right there in your face—and that's why it's so magical.

The more we see the *Falcon*, the more *Star Wars* films there are that feature it, the less we can grasp it, own it, appropriate it, know what it's all about. Now, the labor theory of value is saying that the more you know something, the more you are working on it—in this case with your knowledge. So you "own" the thing you know a lot about—hence the slang "own" meaning "to put down." You've "got it down"—the hint of subjugation in this phrase is accurate. But it turns out that no matter how much you know about the *Falcon*, you are never ready for who's going to fly it next. In one of the "Star Wars stories" prequels, you find out that a feminist robot keen to liberate other robots from being the property of lifeforms can fly the *Falcon*. She's not interested in owning it. When she is "killed" (maybe she really is killed), the remaining crew incorporate her brain into the *Falcon*'s navigation system. The *Falcon* is then really a "she" insofar as the *Falcon* is a feminist robot keen to liberate other robots from their status as slaves—as property that can be destroyed.

The *Falcon* has a feminist, revolutionary navigation system, a system that doesn't regard machines such as droids as inanimate objects that can be exploited. The slavery theme in *Star Wars* is very often coded through droids. "We don't serve your kind here" is something a bartender will often tell a droid: it's all about slavery and segregation. Aristotle said that a slave is "a tool with a soul" (*organon empsychon* in Greek). But in the logic of *Star Wars*, a slave is *a soul that has been enslaved*, reduced to the status of a tool, for someone—but whose basic nature exceeds their functionality and purpose for the master. An

OOO object—"before" (logically before) it is "for" something or someone—is totally real and happy all by itself. That's because there is an aspect of it that is radically impossible to grasp. Slavery is violent because the world really isn't like that. The world isn't a bunch of tools, some happening to have souls, waiting to be used by someone; it is not a bunch of semi-destroyed beings waiting to be taken up in the service of destroying and appropriating other stuff as the property of the master.

This ungraspable quality of things is exactly what the *Falcon* stands for. The *Falcon* appears as the third nonhuman Important Being in the first *Star Wars* film, the one that became *A New Hope*. First we meet C-3PO and R2D2. Then we encounter the *Falcon*. We've become used to machines being people by encountering all the droids. We've become aware of how things can be recycled and repurposed thanks to the droids' adventures being sold by the Jawas on Tatooine. The *Falcon* shares this recycled quality, and the sense of being lost and found, rented out to passengers who want to escape from "Imperial entanglements." The *Falcon* isn't as animate as the droids and it's nowhere near as anthropomorphic. It's truly nonhuman. But it's the first seemingly inanimate thing with a name, and a personality. The *Falcon* is definitely female. Patriarchal discourse refers to craft and ships as *she*. Specifically, the *Falcon* is called *she* plenty of times in the first few *Star Wars* films. By the time the *Falcon* is piloted by the robots-rights quasi-feminist droid L3 and by the woman Rey, this radical potential has become clearer. *Solo* and the third trilogy of *Star Wars* films revises the patriarchal gender tensions of the original films.

The irreducibly hidden, occult aspects of things lead to a consideration of how gender works with the binaries of subject versus object, master versus slave, tool-user versus tool. In OOO, anything could be a tool: a leaf is a tool for an ant. But nothing is just a tool. When so-called tools malfunction we see this clearly. *Star Wars V: The Empire Strikes Back* (1980) is all about a malfunctioning *Falcon*. From the first moment we see it being repaired by a frustrated Chewbacca, to all those moments when the hyperdrive fails, the *Falcon* doesn't do what its pilots require of it.

For OOO, there is no subject versus object. It's all "objects"—which if you prefer could be called entities, because "object" might make you wince and immediately think of something plastic that someone can manipulate. OOO is definitely saying that there are no beings that are "for" being slaves or tools, no beings that intrinsically "look like" they are in slave or tool format. Beings have to be made that way by social relations, power dynamics. There's a lot more to so-called objects than their *objectification*—their being caught in human systems of meaning and value, very often at their peril. Some of these entities are without doubt human—so it's very important again to realize that "object" as I'm using it doesn't imply a (male) (white) (master) "subject" at all. That's the trouble. When we hear the word "object" we see as if in a mirror the very worst thing that could happen to us. We should've called it de-objectification oriented ontology, but I guess we ran out of ink.

And talking of objectification and how to resist it, we are about to explore the most luminously popular image of a de-objectified world, an image that has powerfully feminist and anti-racist resonances.

Let's make the jump to hyperspace.

3 HYPERSPACE

"You wanna *make* that move. You wanna *make* that move." So says Tobias Beckett in *Solo*. He's teaching Chewbacca how to play chess on the holographic board in the circular lounge at the core of the *Falcon*.

But the *Falcon* just *is* a lounge, a lounge that can fly. No wonder Mel Brooks recast the Falcon as a flying Winnebago caravan in the spoof *Spaceballs*.[1] One American fantasy about cars is that they secretly enable you to remain a couch potato while flying across the deserts of the Western USA at a hundred miles per hour. The SUV speaks directly to this fantasy—consider almost any advert for one. Sedan cars built around the time of the first *Star Wars* film were gigantic sofas on wheels, often outfitted in a certain light shade of brown leatherette. Sports cars too, as far as the leatherette went, and in particular the body-hugging seats of the Corvette, are nicely positioned between the feeling of being in an aircraft and the feel of a La-Z-Boy armchair, the schlocky version of the much more stylish Eames armchair. The chair has all kinds of controls that enable you to feel in charge of your letting-go, as you make fine adjustments to its footrest and tilt.

Arriving in the *Falcon* is all about learning how to sit in a leatherette swivel chair. Once you've figured out how to sit in that kind of chair, you might need to go below and swivel about in a much more radically swivel-y chair, the ones from which you fire the rear-facing laser canons. I associate those swivel chairs strongly with 1977, because it's when my grandfather returned from a trip to the USA with a couple of them. My brothers and cousins used to love to spin each other around in them, wondering whether if we span fast enough, the occupant would enter another dimension.

"That's no moon. That's a space station." Obi Wan Kenobi interrupts the quickfire banter with an observation that jolts everyone out of their we're-on-a-cruise-in-a-living-room stance. It's the first *Star Wars* film and the *Falcon* is caught in the tractor beam of the Death Star. The trouble with the Death Star is that, like the *Enterprise*, it's an open-plan office, a place were you work under surveillance and your every reaction is monitored. If you're not sure whether or not the Death Star is an open-plan office, watch British comedian Eddie Izzard's sketch about

Darth Vader in the Death Star canteen. It's very funny because, like all comedy, it's based on an insight we aren't directly aware of.[2]

Lounges are quite different. Lounges allow you to conceal stuff. You aren't automatically under surveillance. The emphasis is leisure rather than labor. Moreover, the *Falcon* is a lounge in which you can travel to another kind of lounge. Sounds cozy, doesn't it? But, in a strange way, this lounge is to be found *inside the Falcon*. Despite this living room's size—it is the size of the universe—you can't get there any other way than by climbing aboard. No wonder when you find it, you sit down in the *Falcon*'s lounge and play chess. Its name may surprise you: this lounge is called *hyperspace*.

In general, we think of space as an empty container, like a three-dimensional map with invisible lines on it. For example, the distance between your face and this page is, you probably imagine, an empty space that is "filled" with things such as dust particles and light rays. But this isn't true. The relativity equations of Albert Einstein tell us that it's a rippling ocean of spacetime. The empty container idea comes from Isaac Newton and it's several hundred years out of date. It's not surprising that this idea has persisted. For one thing, it doesn't seem to matter very much at human scales that we move around in a sort of liquid emanated by physical objects. But there's a deeper reason. In the background is the way humans have confused space and time with the *measurement* of space and time, for thousands of years. One of the foundational moments of "civilization" (the agricultural sort that started about 10,000 BCE) is exactly this confusion of space and time with their measurement.

We think of time as minutes and seconds and years, but those are just measuring categories. We think of space as meters and miles, but those are also just measuring points. This is exactly the same thing as confusing *logic* with *logistics*. To get stuff done, you need an underlying logic about why and how stuff gets done. We don't confuse heat and temperature. We know heat is a kind of radiance: we can feel it when we hold a hot cup of coffee. We don't confuse that feeling with what we see on a thermometer we dip in the coffee. We don't seem to know emotionally that space and time are actually *feelings* or *feels*.

When spacecraft travel faster than light, they enter a realm we often call *hyperspace*. Fascinatingly hyperspace is definitely *not* an empty box. You would have thought that the *hyper* in the word meant that whatever space is, hyperspace is a lot more of it, just a bigger or grander empty box. But it turns out that *hyperspace has a feel*. Hyperspace is "space-feel" just like potato chips have "mouthfeel" in the PR business. (The word *mouthfeel* has really bad mouthfeel.) But there is a much simpler way of putting this:

Hyperspace is a place.

Not only that, it's a really nice place. It should really be called *hyperplace*. This is a truly remarkable phenomenon. It means human beings are capable of rising above (*hyper* means "above" in Greek) the idea that space is a blank neutral container. Humans can come into sync with the developments in philosophy and physics that rocked the world around 1900. These developments are the discovery of spacetime by Einstein, and the discovery of phenomenology by Husserl. What do I mean? Ordinary people realize, in a visceral and experiential way, that we are not living in the Middle Ages anymore.

It's not just about science, it's about culture in general. Spacetime (general relativity theory) and phenomenology might be very much the same kinds of development, and turn-of-the-century art is an indication. Only consider Claude Monet's water lily paintings, most vividly and accurately displayed in the Orangerie in Paris. The Orangerie is a pair of oval-shaped rooms in the Jardin des Tuileries. An oval is a circle that's been squeezed, just like spacetime isn't totally regular but is squeezed and stretched by gravity. An oval is also an egg, a vessel that is both a container and a living habitat for an embryo. The visitors are the embryos in the egg of the Orangerie, and Monet's paintings are the yolk, a gorgeous, mauve-blue-green yolk. Floating in the yolk are little blobs, the water lilies. They appear not as objects in empty space, but as an intrinsic part of their habitat, the slowly rippling, smooth, transparent liquid of Monet's pond at Giverny. The pond contains so much else—water weeds, shadows, the sky. The way Monet paints them, it is as if the lilies emerge out of this medium. They don't just sit in it or on it, it is as if they are symptoms of the water itself, growing out of the mauve or blue or green liquid.

This is just like how Einstein imagines spacetime. It's also just like how some experimental novelists imagined prose. When I teach my how-to-read-literature classes, I start with the idea that prose is really poetry with really, really long lines that have to be right-justified and called paragraphs (nowadays anyway). It seems easy and helpful to do so. But at some stage there's a flip, when I determine the students are ready for it, and instead of teaching prose as really long poetry, I teach prose as the overall medium which sometimes can become "hotter" or more "lumpy" or "reified" as poems. On this view, poems are little islands like planets or water lilies in a more general liquid like a pond or spacetime. "Prose" is the basic stuff. Water is the basic stuff. Spacetime is the basic stuff. Hyperspace is the basic stuff—*hyper* no longer in the sense of beyond, but in the sense of concentrated or condensed, like condensed milk or consommé.

There was a special kind of prose that arose in English literature around 1900, the kind we call stream of consciousness. We associate it most with the work of Virginia Woolf and James Joyce. Of the two of them, Woolf is the most radical. Her streams of consciousness appear wide enough to contain more than one person's mind, as if everyone were capable of flowing into everyone else through some kind of telepathy. Woolf's genius was to push a logical consequence of narrative realism to its limit. Realism, as she herself argued, isn't about detailed descriptions of things. It's about a *feeling* of reality. Through a brilliant narrative technology that Jane Austen developed (it's called *untagged indirect speech* and it would sadly be a digression to describe it in detail), the reader can have something like a telepathic-seeming experience with a totally unreal person, a fictional character. In this way, all realist stories are ghost stories. But if you can get inside one character's head, why not several, all at once? And since "mind" or "soul" isn't really a kind of smoke trapped in a bottle (the soul or mind versus body dualism of Medieval patriarchy), why not?

The point is, stream of consciousness shows you something very interesting. It shows you that along with thoughts, there's an entity that is having the thoughts, in which the thoughts are appearing. It's not so much what you're thinking, but the fact that you're thinking it. The fact that your thoughts are taking place "within" a liquid-like consciousness.

Hyperspace is in a way regular old space, but imagined as a *phenomenon* along the lines of Husserl—as a *thing* with specific characteristics, not just a blank slate. Space is supposed to be what contains all other phenomena. Space itself isn't different from other phenomena such as planets and people reading a book called *Spacecraft*. That means it's not omnipresent, and that omniscience and omnipotence are therefore impossible. Space isn't evenly always the same. It has contours. It is creamy. The way hyperspace is visualized as kind of roaring blue and white pearlescent ocean is a wonderful sign of real political, intellectual and artistic progress. The creamy phenomenology of hyperspace means that space is palpable, not different from galaxies and gnus. It means that the "whole" of the universe is one thing, the same as its parts (black holes and boots), so that the whole is amazingly "less" than the sum of its parts. It's not a scary infinite mouth at all. It's not swallowing everything and being ontologically bigger than everything. It's a swirl with blobs called quasars and cockatoos.

Hyperspace is the *ecological niche* of spacecraft—their *habitat*. Hyperspace independent of them. Hyperspace is there whether or not you manage to attain it. A spacecraft enters hyperspace very suddenly, like a marble plopped into a perfectly marble-shaped hole. If the spacecraft is the baby, then hyperspace is the milky bathwater. They go together like a terrapin and mangrove forest. And just like that mangrove forest, hyperspace isn't empty and dark and cold. It's warm and luminous.

The term *hyperspace* was first coined in 1867.[3] Thirty years later, the term had evolved to suggest the two essential ingredients of the geometry of the luminous ambient realm I am discussing here. First, *hyperspace* can refer to any space of more than three dimensions. Spacetime is four-dimensional since time is treated as a fourth dimension. Secondly, *hyperspace* can refer to non-Euclidean space. The mathematician Carl Friedrich Gauss had figured out how to create an entirely new geometry, based on curved surfaces. If you draw two parallel lines on a piece of paper, they will never meet—they will "meet at infinity" as the Euclidean saying goes. If, however, you take a felt tip pen and draw two parallel lines on an orange, they will eventually meet—at the navel.

This is Gaussian geometry. It was pivotal in Einstein's description of spacetime. To get the notion of squishy, rounded spacetime across, his text charmingly refers to "reference mollusks," as if spacetime were made of mussels and oysters.[4] The aliveness and wetness of these images is what hyperspace is all about. Spacetime isn't like Newtonian space at all, frigid and dry. Spacetime is damp, and hyperspace is also luminous and glistening.

"Traveling in hyperspace ain't like dusting crops, boy." Han Solo says to Luke Skywalker in *A New Hope* before the first time the *Falcon* disappears into it. At other moments it is said that the Falcon "makes" the jump to hyperspace. "Making" sounds a little like how the verb is used in "making love"— achievement mixed with creation, nicely poised in the middle of the patriarchal binary of active and passive. Making hyperspace is thus far, far different from "dusting" a flat Euclidean and Cartesian surface with killer chemicals. In that case, there is a subject and an object, an active and a passive, a master and a slave. There is contingency involved: you might get it wrong. Han is giving Luke a valuable lesson that we could extend to how to grow up from the oppressive medieval Neoplatonic programming.

Han's line shows that the Jedi are not the fount of all wisdom in *Star Wars*. Far from it. Their original sin seems precisely to be all about trying to establish a medieval, patriarchal hierarchy (the "rule of the priesthood"). This doesn't mean Han is saying something perfect. All kinds of heteronormative anxieties are tangled up in what Han is saying: he is an older man displaying his prowess. But he shouldn't worry. In *Solo*, we see that Lando Calrissian has buried the memory of L3-37, his feminist robot girlfriend, deep in the guidance system of the *Falcon*. The males tinkering with the knobs in the cockpit shouldn't be so concerned—females are in charge.

When the *Falcon* finally "makes" hyperspace, it is as if swallowed in the vagina—or for that matter the anal sphincter—of its substance. The baleen of the black-and-white lines of stars and space seems to engulf the *Falcon* and to twist slightly like a sphincter opening and closing (think of an iris or a camera shutter). The *Falcon* is often about to get swallowed by other beings—a "gravity well" or black hole, a gigantic space monster rather like a sea-monster, a huge worm that bites as the ship escapes its stomach. The only tube it wants to be swallowed by is hyperspace, thank you very much. This is the tube of staying alive, rather than of dying.

Star Wars is all about the Force. In short, *Star Wars* is all about telekinesis, or as Rey puts it, "lifting rocks" (without touching them). One of the primordial features of the *Falcon* is its effortless ability to defy gravity. People push gurneys around corridors. Land speeders float across the surface of a desert planet. But the *Falcon* is the first vessel we see taking off from a planet and exiting its gravitational field. The *Falcon* weaves together the mundane world of weirdly snouted aliens in bars and the spacious voids of the far-away galaxy, haunted by the ships of the empire, which tend to lumber, or just sit there (The Death Star), or appear with a sudden horrifying jolt and come to a full stop. The *Falcon* weaves. The language of anti-gravity and hyperspace travel has very much to do with cloth. We talk about how wormholes, if they exist, fold space in an extra dimension, enabling travel between unimaginable distances. In *Star Wars*, the *Falcon* is the craft that introduces us to hyperspace.

We never really see warp space in *Star Trek*—we see ships zooming into warp drive and there are some scenes where we witness hyperspatial stuff outside a window. There is the infamous scene in the first *Star Trek* movie (*Star Trek: The Motion Picture*, a scene that took weeks to shoot), in which the Enterprise creates an artificial wormhole. And there's a scene in the second reboot (*Star Trek: Into Darkness*) in which a heavily weaponized star ship pursues the Enterprise, and we see the "tunnel."

In *Star Trek* there are degrees of warp (factors one to ten), whereas in *Star Wars* there's just hyperspace or not hyperspace. The idea that you can control how much warp you have is deeply technocratic and even imperialist. If you can have degrees of warp, then you can measure it. Whoever has the most accurate way of measuring warp has the most power in the *Star Trek* universe. But you can't measure hyperspace; it's not a thing to be measured. It's the whole idea of measuring anything at all—what the Kantian sublime evokes.[5] The Empire can therefore never have a monopoly on it. They can't parcel it out. They can't own it. *Hyperspace is a thing*, a thing in its own right. The seemingly democratic *Star Trek* loses to the seemingly reactionary *Star Wars*.

This difference between warp and hyperspace is a bit like the political and philosophical difference between Hegel and Kant. For Hegel, there are amounts of realization, which ends with him saying that white western people have more of it than others. In Kant, there's just freedom and not freedom. It's like hyperspace. Either you can make it, or you can't. It's not like dusting crops because it doesn't imply an agricultural hierarchy in which there are landlords and peasants and slaves and plants. There's no gradation. So "making" it always involves a kind of luck or serendipity. You can't make it a little bit, then get better at it. You just set the controls and zoom. You can't have more of it, once you've made it. It's not a thing you can parcel out. Its thingliness is incarnated on the screen as a luscious black- or blue-and-silver fluid, as if the wonder of the early silver screen were invading the mundane glitziness of technicolor.

Hyperspace settles down from the initial whoosh to a whirl or indeed a *whorl*. The curves of a spiral shell; the patterns of fingerprints; the radiation of leaves and flowers from a stem; a convolution or wreath suggesting movement—*whorl* captures very well that quality of moving while standing still, and a quality of mathematical beauty, *and* a quality of being a physical, substantial *thing*. A whorl on a spinning wheel is a small fly-wheel that regulates the wheel's speed.[6] A sense of containment, of order, of beauty . . . hyperspace is not just the sublime freedom rush of the whoosh. My idea of *hyperobjects* comes in handy here. A hyperobject is something so vast in space and time that you can only "see" bits of it at once, and yet it's not infinite. It's just really really large—our biosphere would be a good example.[7] Hyperspace, as a hyperobject, has a kind of *finitude*—it's this way, it's not that way, it has *this* kind of movement, not *that* kind. In lots of ways actual hyperspace is a gigantically beautiful entity, not a sublime infinity.

Hyperspace or warp space has been a tunnel since *2001* and *Doctor Who*. Stanley Kubrick lived in the UK and must have seen the way the opening credits evolved to pay homage to his *2001* Stargate. What is the basic phenomenology of traveling down a tunnel? It is moving while being still. The end of the tunnel may grow larger, but doesn't arrive; the "end" of hyperspace never seems to grow larger. At a distance, it might grow very slowly from what appears to be a small point—the light at the end of it. (Many people who have had near death experiences (NDEs), or who have taken DMT, or studied the *Tibetan Book of the Dead*, report about this tunnel.) And the tunnel has its own reality—it's not just a dark empty tunnel. It is luminous and flowing, and in the case of the mid-1970s *Doctor Who* tunnel, it appears to be infinitesimally ribbed.

I'm talking here about sex not merely to titillate, but to make a political point. Hyperspace is pleasurable, blissful. There's a reason for all those disco references to hyperspace.[8] After watching *Star Wars*, the meditation students of Chögyam Trungpa would refer to the states of meditation bliss as "hyperbliss."[9] Hyperspace is about what Blake called "an improvement in sensual enjoyment."[10] Spacecraft "making hyperspace" has to do with achieving a utopian energy, perhaps an energy not based on the deadly burning of fossil fuels, a more benevolent form of human coexistence with one another and with other lifeforms.

The "howlaround" effect in the opening credits of the early series of *Doctor Who* also conveyed feelings of a rippling, flowing movement-while-still. It's quite an easy special effect to achieve.[11] Significantly, the use of the word "howl" captures, as if synesthetically, the "sound" of rushing through a tunnel rendered as a visual effect. NDEs often include very loud sounds, like the churning of the *Doctor Who* TARDIS or the sound of hyperspace in Star Wars, or indeed the rushing sounds in Kubrick's Stargate sequence. Or, as one Tibetan teacher says of the "in between" tunnel of the bardo between lives, "the roaring sound of the immeasurable void."[12]

The hyperspace of *Star Wars* is a fully, sensually realized time liquid, a liquid that influenced the opening credits of the *Doctor Who* reboots since 2005. It has nothing to do with the after-death bardo space that *2001* evokes. It's trippy, but that doesn't mean it's mental or conceptual or psychological. It has a physical feel to it. You enter suddenly when you push forward a certain throttle on the control panel. The stars ahead begin to stream light as if points are turning into lines. Then all of a sudden, zoom, there you are.

What happens is well described by the feminist word *circlusion*.[13] We often think of spacecraft, for obvious patriarchal reasons, as phallic symbols penetrating the hyperspace tunnel. But it is more accurate, and less violent, to think of it the other way around. Bini Adamczak developed the very helpful verb *circlude* and the noun *circlusion* to describe any process of enveloping—a hand around a sex toy, or an anus around a finger, for example, or a mouth around a nipple. Here's a salient paragraph:

> Think of the moment when you were taught in school how to prevent the spread of sexually transmitted infections. No one would ever think of trying to push the banana into the freshly unwrapped condom, would they? But the task of correctly applying a condom is easy when you think of it as unrolling the tube onto the banana. Indeed, circlusion is an extremely common experience of everyday life. Think of how a net catches fish, how gums envelop their food, how a nutcracker crunches nuts, or how a hand encircles a joystick, a bottle of beer . . .

Adamczak coaxes us out of the habit of associating penetration with certain parts of the body and with a certain kind of violence. Hyperspace *circludes* the spacecraft. The roaring, wheezing sound of the TARDIS as the Doctor enters hyperspace is something like the grinding churn of air or fluid surrounding the craft and entering into some kind of friction with it. (It was created in the BBC Radiophonic Workshop by dragging some keys across some piano strings.) Indeed, at one point the Doctor is told that he is basically leaving the parking brake on and that the ride through hyperspace could be much more silent and well lubricated. The way the Falcon appears to spiral or swivel into this circluding liquid reminds me of the leatherette swivel chairs in the various *Falcon* cockpits, and that one of the original senses of *swivel* is *fuck*.[14]

Hyperspace *circludes* the craft because, at light speed or faster, that craft isn't really traveling at all. This Einsteinian factoid is suggested by the engineer, Scotty, in the first *Star Trek* reboot, when a very old Mr. Spock types out his equation for trans-warp beaming, teleportation at speeds faster than light. "Imagine that—it never occurred to me to think of space as the thing that was moving!" It's why I very much like this book's cover design. Hyperspace on the cover of this book is its own "thing," a place that seems in fact to be approaching the ship, rather than the other way around. There is a tension between the supposed reality of the Cartesian and patriarchal space that the ship is traveling through, and the hyperspace that is about to envelop it.

We are moving from a regime of penetration to one of *circlusion*. In the *Star Trek* reboots, the spring-like recoil is definitely as if space itself were sucking the ships into it—circluding them in its warp, a word that suggests the feminized activity of weaving and its soft, winding, twisting fabrics, rather than the straight and narrow grids of Cartesian space. In the *Star Wars* series, when a vessel enters hyperspace, they disappear, exactly as if they had been enveloped by a thick medium that makes them invisible. Warping, the thing that the Norns do, the Norse female beings entwining us in the web of fate.

The word *weird* is derived from the old Norse *urthr*, which means *twisted into a loop*.[15] Weirdness, "fate," and the deja-vu-like feeling that it's happening, is the twisting of twine into a loop—the thing western civilization has been trying to straighten out into "fate" as a matter of fact. Fate is set in stone and mechanical, you can't stop it. Weirdness is twisted and strange and knowing. Warp drive, metaphorically speaking, requires weirdness.

The *Falcon* is many things—a Frisbee, a super flat sports car, probably a Corvette, the sort of one that an American working-class person might be able to afford, just about, if it was second-hand. It's a sand dollar. And it's also a vulva—a vulva rushing through the vulva-like realm of hyperspace. It's what in heraldry and in deconstruction is called a *mise-en-abyme*, where in a shield design there is a little version of the shield design in the middle. Who or what is "making" whom or what? The vulva of hyperspace circludes the vulva of the *Falcon* which circludes the passenger pilots . . . I think that the reason the political right appropriated *Star Wars* is because it *doesn't* really fit their agenda.

The evocation is of *something beyond speed*, a speed beyond speed itself, as if we were experiencing velocity as such, rather than a specific one. It's like what Kant says about *the sublime*. The sublime gives you a feel of magnitude, beyond any specific number or amount. Try counting up to infinity. You never get there . . . then you suddenly realize, whoa, *that is infinity*—a number I can't reach. It's not a series of nines going on forever. It's uncountable . . . it's a kind of number, but you don't know what.[16] Infinity is the withdrawal aspect of an OOO object. You can literally hold it in the palm of your hand, as in the Blake poem—it *is* the palm of your hand.[17] Infinity means there might be three of whatever it is, but you can't know yet. This not-yet quality, this futural, open quality is exactly what we are experiencing in the sublime, and if OOO is correct, it is a basic feature of every entity in the universe, from spoons to Grand Moff Tarkin. Everything is a TARDIS—everything is "bigger on the inside"—infinitely bigger. That's a thing I said in my book *Hyperobjects*, and as I observed just now, the hyperspace of *Star Wars* is a wonderful example of a hyperobject: something so massively distributed in time and space—it sort of just *is* time and space, everywhere and everywhen all at once—that you can't point to it. It's invisible because it's everywhere.[18]

That's like the future. The future is everywhere, you can reach it any time—the real future or as I like to say sometimes, the *future future*. Futurality is the possibility that things could be different. It's like what Blake says, again, about seeing eternity in a flower, or how Walter Benjamin talks about how each moment could be a gate into utopia. It's not visible, but you can feel it. The hyperspace of *Star Wars* is the closest a regular cinema-goer gets to realizing that film isn't just about the visible, but about the invisible, and in particular, the feel of motion, all that liquid celluloid spooling through a projector (*Star Wars* was shot in luxurious 70mm). It's a lovely gift that George Lucas gave to ordinary people—something utopian to keep them going.

I dislike the popular left-wing cynicism and despair that ideology is everywhere and that my telling you how paralyzed and trapped we are indicates how much more intelligent than you I am. I believe that the "ideology is everywhere" ideology is wrong for the right reasons. It's true, but it misses something crucial. Ideology may be *everywhere*, but it's not *everything*. Another way of saying this is that the past is everywhere—just look at the wreckage around you we call "the present moment." Everything is a story about how something happened to that thing. My face is a map of all the acne I had when I was nineteen years old. But who is this Tim Morton anyway? What exactly is a face? What does this poem—which consists of hundreds of deliberate and non-deliberate decisions made to put a rhyme here, an image there— mean? What is this building? We can see how it was made by looking at it and walking around it. But what is it, really? This window isn't just "for" someone to look through. It's also a landing strip for dust and flies. So, what is it?

The past may be everywhere, but it's not everything. Things could always be different. Everything has a "hyperspace" quality. To this extent, it really is perfectly possible to "see" the future. In *Star Wars*, this seeing is more of a feeling—a recoil.

Everything has a secret passage, a hidden corner, because of time. When I round the corner in this hotel in Brooklyn I've just arrived at, searching in vain for my room, the corridor I was just walking down becomes the strange, hidden corridor. The secret quality of things is the future. So it's not totally hidden: you can feel it. We can actually "hear" a secret. Try it. Say to yourself, "Some things are impossible to speak about." You just *did* speak about them. It's like leaving Plato's cave, which is a great proto-cinema. When you leave, you're so blinded by the light that you can't see it. Then you go back down into the cinema-cave and try to explain, but no one else has "seen" this blinding light, so you can't say it. It's like you can taste sugar but you can't speak, which is the analogy Buddhists use to describe the elusive but real (to them) "nature of mind." The future isn't visible, not visible in an "ocularcentric" objectified way, there before you look, but you can feel it in a sensual rush, a physical movement of orientation. That's what the "whoosh" of making hyperspace is.

Something crucial occurs before you can leave Plato's cave and see the truth. You must twist or turn around in your seat. That's exactly what making hyperspace is. It's a twist or turn like that, profoundly kinesthetic. This kind of turn was noted by Martin Heidegger in his wonderful and spooky reading of Plato's cave.[19] When we talk about a "linguistic turn" or a "nonhuman turn" and so on, we are using Heidegger's language. To see, you first have to orient.

That's what we're seeing when we enter hyperspace. We're "seeing" withdrawal. We're seeing the outside of Plato's cinema cave, or rather feeling it in the basic twisting turn, like trying to see something out of the corner of your eye—when you look directly, it's gone. It's not like the private experience of Dave in *2001*. We see something through Dave's eyes, but we don't see the blinding flash of the future itself. We see a bunch of trippy colors and patterns. It's more like the "room" in Tarkovsky's *Stalker*.[20] We "see" a room in which all our wishes could come true. Trouble is, that idea is paralyzing. Yes, I'm saying that *Star Wars* is more sophisticated than a Kubrick film, more philosophically sophisticated, and more politically sophisticated, and readily available to anyone who cares to watch a YouTube clip. And I might even be saying that it's more politically sophisticated than a Tarkovsky film, sacrilegious horror of horrors. That's because hyperspace isn't paralyzing, like the room in the Zone in *Stalker*. It's not like the alien idea of god (Tarkovsky plays with Christian themes a lot). Hyperspace doesn't freeze you in the headlights. It's the very possibility of moving at all, anywhere. And anyone can achieve it. It's not for an elite selected by the military like in *2001* or people who pay for an exclusive tour like in *Stalker*. You can "make" hyperspace even if you're a gun-slinging low-life from a seedy bar in a dirty desert town.

In *Star Wars*, we are not alienated from the possibility that things could be different. We are not alienated from the future. Fascism believes we have been alienated from a "great" past that we could somehow find underneath the scraps and fragments of the present. I think that socialism or communism believes we have been alienated from the future. We don't say it much. We often say that the idea of alienation implies some kind of authentic state "from" which we have been excluded (the past). But this is just religion. It's *right for the wrong reasons*—we are alienated, but not from the past. Once you get used to it, the idea that we have been alienated from the future is much easier than thinking about some kind of fall. Exactly how could a fall have happened if everything was perfect? And just look around you—if you can see stuff that isn't quite right, it means that the past sucked. It couldn't have been perfect. The past is just all the crumbs of all the cookies that crumbled so far. But all the mountains of cookie crumbs in all world don't imply that there isn't cookie dough ice cream.

The sophisticated "ideology is everywhere" theory—the thing someone like me is supposed to teach you in theory class is, as I just said, *wrong for the right reasons*. The crude "we are alienated from a natural state" theory is *right for the wrong reasons*. We are alienated from the future, which is strictly invisible and unspeakable, if you try to point to it with a spotlight, but not "un-feelable." That's the other thing about *Star Wars* hyperspace, talking of spotlights. It's as if there's nothing to focus on. It's just a flood of light. We are illuminated by the cinema screen, which has become a floodlight.

It's like the end of a rock concert when the lights turn on the audience. It means "You've got the controls, you're part of this, everything you see on the screen, all the characters and the action, is the holographic R2D2 projection of you, the audience." We often forget that the audience is part of the drama because of the conventional fourth wall that turns the theater into a shop window display or fishbowl that we are peering into. But drama comes out of dancing—that's what "chorus" means: singing, dancing, "raving" people. Sometimes someone steps out and does some moves, and then this gets formalized as "the protagonist." Then someone else steps out (the antagonist, the supporting characters) and so on. Eventually the chorus just evolves into the commentator or the laugh track and we forget that they are the whole thing, and that we are part of the chorus.

That's also how to understand a pop song. A lyric is always about the chorus. The verses are just "examples" of what the chorus is talking about. It's not that the chorus is a commentary on the verses. It's the other way around. This is Nietzsche's theory of how tragedy arose in ancient Greece.[21] It is now speculated that language itself, as in words and phrases (loads of beings "have" language in other ways), emerged out of rituals that were like humanoid primates raving and going "make some noise!" as one might do in a nightclub.[22] Language came out of applauding.

No wonder there was a round of applause when the *Falcon* first "made" hyperspace. It's not just because it's a dazzling effect. It's because it's a fourth-wall-collapsing, deeply empowering, you've-got-the-controls moment. We all become pilots of the *Falcon* at that moment. The *Falcon* making hyperspace is a moment of un-alienation. It's realizing, halfway through a very immersive drama, that there is an outside of the cinema, that the reason this is all happening is because you paid for a ticket, and because George Lucas made a film with the idea of you wanting to see it very much in mind. It's a moment of release from the pesky Imperial cruisers around Tatooine and it's just a stunning visual effect, in which the cinema screen is flooded with bright lines, as if it had turned into a Bridget Riley painting, rather like how the sounds of hyperspace and the TARDIS resemble the found sound and *musique concrète*. A representational, cinematically realist film gives way to something by Stan Brakhage, just for a second.

Hyperspace shows the cinema viewers that they are looking at a screen. It's a dazzling effect of being undazzled by all the characters and stories. It's like that Situationist adage, "Beneath the street, the beach." It's a deeply anti-imperialist moment. The statues and the concrete may be all around you, oppressing you. But really there's nothing stopping you from tearing them up and chucking them in the river. Hyperspace is that river.

This is where we need to talk about the dominant visual effect for producing hyperspace: slit-scan.[23] John Whitney pioneered it for *Vertigo*, and Douglas Trumbull saw the sequence that first uses it. Trumbull subsequently made the effects for *To the Moon and Beyond*.[24] Stanley Kubrick saw this film and employed Trumbull to make the Stargate sequence of *2001*. The effect was also pioneered by Bernard Lodge for *Doctor Who*. Slit-scan was still used in some way to create the hyperspatial tesseract scene inside the black hole toward the end of *Interstellar*.

Slit-scan involves time, adding an extra sliding or gliding through time to produce the image. The medium through which the film is shot—the slit, is the thing that moves: "It never occurred to me to consider space as the thing that was moving." Slit-scan also implies immersive, palpable space. Panoramic photography was the first technology to employ slit-scan. Before they were photographs, panoramas were environments, huge cylinders that one could walk through, descending a spiral staircase. You move and the immersive painting moves with you. You, in a sense, are the moving slit. There were panoramas in Leicester Square in the Romantic period, and William Wordsworth visited them. It's quite possible that this deliciously popular experience is what enabled him to invent a whole new poetics that destroyed the objectification and fixity of the "landscape" art of the picturesque.

Picturesque art is still with us—it's called the selfie. Back in Wordsworth's time, the selfie was an inverted one—a charming image of a landscape (a word that itself means an *image* of land) that displays the good taste of the viewer of painter. Before cameras showed up—before smartphones with their fish-eye lenses reinvented all this for us—people used the Claude Glass. It was a hemispherical lozenge of sepia-colored glass that would turn the view into an aestheticized image, albeit upside down, sepia-tinted like a tasteful ink painting. To achieve the best effect, users of the Claude Glass would get into a position called *repoussoir*, down and to the side of the view. This is the exact inverse of the up-and-to-the-side view popular for selfies today. And that's because it's really the exact same thing: it's just a selfie made of nonhuman beings. That's the trouble with landscapes, including landscape architecture, which does it for real: they're selfies made of nonhumans, like Archimboldo paintings.

Not so the panorama. And this counter-tradition of immersive, disorienting popular art continued into the twentieth century. *To the Moon and Beyond* was projected onto a domed screen, which must have made Trumbull's slit-scan effects particularly immersive for Stanley Kubrick, who was sitting in the audience as the film played at the World's Fair in 1964. John Whitney's experiments had to do with spiraling—a moving and opening while standing still. His spiral opening out of the eye in *Vertigo*'s opening sequence is the classic instance. The technique has some resemblances with the pullfocus or "dolly zoom" which Trumbull also used for the Stargate sequence. In a pull focus, a camera zooms in on an image while being pulled away from that image. The effect is uncanny, as if a thing we might experience as static is somehow moving while not moving, as if breathing with a life of its own.

This is precisely the living space of hyperspace, a breathing, palpable thing. Organic, yet not necessarily "alive" versus "dead," and so uncanny. That kind of "alive" is overrated anyway—it's what we call survival mode, living as resisting death at all costs, which Freud brilliantly called the death drive.[25] Life as such is more like a quivering between two different types of death, a slit of flesh moving between not existing at all (decay) and doing the same thing over and over again (burning out). Life is the sound and texture and look and feel of rippling flesh.

The best way to think of an OOO object is to think of it as a liquid. Quick, think of an "object"—you visualized something solid didn't you, like a ping pong ball or a LEGO brick. In particular, hyperspace has a body-fluid-like liquidity, so it seems vital to rethink what we mean by "object" to come to terms with it. Hyperspace moves and slips around itself, all by itself. The first female Doctor Who—the latest incarnation, played by Jodi Whittaker—pilots her TARDIS through a kaleidoscopic, substantially viscous version of this time liquid. Spacetime as a liquid appears in the dimensional portals in the beautiful Afrofuturist short film *Until the Quiet Comes*, by Khalil Joseph.[26] They are utopian portals that allow time to flow backwards, reanimating the body of a murdered Black man, whose beautiful, uncanny backwards-seeming dance evokes a feeling of time being capable of redemption. The Kalunga, the Kongo portal between the worlds, is the watery liquid that we see swirling at the beginning, middle and end of the film, and this Kalunga is exactly hyperspace.

When we enter or "make" hyperspace, it's also like having an orgasm. The fourth wall collapse I discussed a moment ago is a kind of orgasm of the film stock, a sudden shudder or "petit mort" as the French say about coming: *la mort, l'amour*. There is a sudden spring or ripple that demonstrates how not in control of our bodies we really are. We move from extreme pleasure into bliss. In the esoteric "tantric" parts of Hinduism and Buddhism, bliss is a pathway beyond the ego— into the hyperspace of enlightenment. Remember that Buddha became enlightened because he ate something creamy and delicious and sweet, like hyperspace—a bowl of rice pudding. Hyperspace is an *erotic* version of space rather than a scary or blank one. In *Interstellar*, the voyage of the *Endurance* through the wormhole seems scary, a rushing darkness full of the clattering sounds of malfunctioning ship mechanisms. But after Cooper discovers how to work with gravity inside the tesseract in the black hole, he finds himself in the hyperspace mother-of-pearl. It's a special five-dimensional cream, all silvers and blacks, in which he can swim and reach into the ship carrying him and his colleagues through a wormhole in the past, to shake hands with his prospective girlfriend, Doctor Emilia Brand.

The return of Cooper through the wormhole to Saturn is through a region that we no longer see as a scary darkness, but is inverted as a white sparkling brilliance, a sort of ice cream sauce, liquid and oozing with honey-colored and dark caramelized lumps.

Cooper's dissolution into the hyperspace sauce, while the geometrical tesseract in which he figures out how to tell his daughter how to discover quantum gravity is collapsing, is remarkably like a religious experience or death or orgasm: "What happens now?" he gasps as he dissolves into light. *What happens now*—it's an orientation toward the future as truly unknown and unknowable, rather than just mechanically repeating the past. *What happens now* is the feeling of falling in love, having a spontaneous twinge such as an orgasm . . . or entering hyperspace.

The encounters of the *Falcon* are deeply physical. In *Solo*, the *Falcon* avoids the Scylla of a gigantic space monster and the Charybdis of a black hole by making hyperspace just in time. It's making the "Kessel run" for "spice," whatever that is. (One perhaps assumes after James Herbert that it's a kind of drug. It's expensive enough.) It's working hard. After a particularly supreme effort, the crew can relax. Or consider the moment in The *Empire Strikes Back* when the characters realize that what they took to be a cave is in fact the esophagus of a gigantic space worm. In *The Force Awakens,* Han makes hyperspace while the *Falcon* is stuck to a giant murderous slimy sphere called a Rathtar.

To the outside world, you disappear when you enter hyperspace. But for you, it's as if you've stepped out of the street into the lounge. *Star Wars* sensualizes hyperspace in this way—hyperspace is not a place of cosmic speed and massive realizations, but rather a quiet lounge. There's an achieved privacy. It's soft and comfy. You can play chess. You can flirt. You can talk about boyfriends. Suddenly you arrive at your destination and your spacecraft floats a little. It's as if you have been in a traveling lounge that can relocate anywhere. Then you have to get to work and do stuff like fighting or landing.

This is a distinct improvement on how hyperspace happens in *2001*. There, hyperspace is a luxury palace not unlike Versailles, and it has a grid on the floor that disappointingly turns it into an empty box. Dave voyages through the spatiotemporal liquid in a coracle-like EVA. We see glimpses of his freaked-out face rushing through the trippy tunnel, and this plus his helmet plus the spherical EVA are not unlike like Scottie's head traveling down the tunnel of misogynistic paranoia in *Vertigo*.[27] Dave voyages through the mystical liquid, complete with sperm and eggs and melting lava, only to find himself in a luxury version of Newtonian physics. Just as *comfort* is more progressive than *luxury*, the hyperspace of *Star Wars* is more progressive than that of *2001*. It's not about becoming some kind of monarch—the cosmic baby at the end of *2001* is surely a super-powerful being. It's actually more like what Nietzsche was genuinely after in his idea of the superman—not someone with six-pack abs, but someone who can play, like Chewbacca at the holographic chessboard. That's ironic, and scary, because *2001* deploys Richard Strauss' infamous *Thus Spake Zarathustra* opening, in exactly the same way in which the Nazi Party confused the superman with what Nietzsche calls the "ultimate man." But *Star Wars* doesn't celebrate ultimate men plotting world domination in palaces—those are the bad guys, the Sith. *Star Wars* is about all kinds of lifeforms fumbling about in a rickety but comfy freighter. Like a nice lounge stocked with board games, hyperspace isn't about marking a difference between classes. Anyone can sit on your sofa, at least hypothetically. "Make yourself comfortable" isn't something you can say about a hard, golden chair.

The lounge is where you sit when you're not at work. So, hyperspace is also about leisure time. It's about the time we think of as in between work shifts. That's because our economic setup is about "working to live." But as one meditation instructor said, when his bored students asked when the break would be, "this *is* the break."[28] Simply push a throttle, and you can arrive at the weekend. Inside of every weekday is a Saturday. It's definitely not a day of worship. It's a nice demotic Saturday where you can kiss Princess Leia and argue about holographic chess and pontificate about The Force. This "moving while standing still" (and vice versa) sensation is the feeling of being in the middle of a story. It's as if you can access the middle from anywhere in the story. The middle of a story isn't the mathematical middle. It's a feeling. This fact that time is in fact a "feel"—that time is *temporality*, the "ality" part of that word being the tail that wags the dog—is what hyperspace makes palpable.

It's also the feeling of being in a disco. Black and gay people felt a lot safer when they made it off the streets in the 1980s into a club where they could dance to *house* music. It's why it's called that. So it's not surprising that hyperspace (and starship troopers, motherships and UFOs) are common features of disco and techno music. This isn't about speed, as a matter of fact. It's about a luxurious, utopian sigh of relief, made of movement, a house made of dancing. All houses are kind of dances anyway and house music just makes that obvious. One almost wants "We Are Family" to be playing as Rey and Finn and Poe and the others get together in *Star Wars 8: The Last Jedi*.

Beneath the street, the beach. Beneath the seeming linear flow of time, which is always some imperialist or colonialist imposition, is the timeless bath of hyperspace. Let's think again about that verb "make." It's halfway between active and passive—building or constructing something, and catching or hopping on: "I made the bus in seconds flat" (Paul McCartney, "A Day in the Life").[29] When you make something in this more passive sense, it's about arriving somewhere and experiencing relief. It's not about mastery. It's about being able to drop the tool, to take your foot off the gas. It's about peace. To this extent, making hyperspace is all about dissolving the metaphysical difference between active and passive, male and female, work and leisure, outside and inside, Monday and Saturday, work and life . . . Making hyperspace is about making peace.

Peace is better than being right or being saved. Peace might be the ultimate image of utopia, as Theodor Adorno once said.[30] Let's go back to how *Star Wars* hyperspace is better than the room in *Stalker*'s Zone because it's not alienating. The hyperspace of *Star Wars* is un-alienation itself, cheaply available in a movie full of cheap thrills. The fact that Han Solo, the first *Falcon* pilot we see, makes hyperspace, is connected to how Luke and Obi Wan can use The Force. This seems rather extraordinary given Han's cynical reluctance to acknowledge the existence of The Force. But in a way, he's right—Han Solo is questioning the religious hierarchy that has grown up around The Force. It's in the tone of his remark, rather like how someone might scoff in a cathedral about the stained glass: all those colors doesn't mean you guys invented color. That's a hard-won insight. The whole point of cathedrals is to make you think that the church has a monopoly on beautiful colors and powerful sounds. (The first hyperspace shot resembles a black-and-white version of one of the rose windows in Notre Dame.)

Star Wars is a truly pagan, non-theistic work of art insofar as there is no religion, in the sense of good versus evil. Anakin could not have become Vader if that was the case. Anakin is able to "cross to the dark side" because The Force is just one thing. The so-called dark side is simply a higher amplitude version of the good old Force, an amplitude that the humanoids in the galaxy far away can't handle. What is wrong with the Jedi is their religious approach to this basic energy, which blinds them and makes them proud and exclusive and hierarchical (Greek, *hierarchy*, rule by the priestly caste).

That's the deep reason why Han Solo can make hyperspace. It's a truly classless galaxy at its core, despite how people have messed it up, because there's no hierarchy in The Force. It's a continuum. It's just that the humanoids can't turn it all the way up to 11 without getting burned. Even the Jedi are getting it wrong, as Luke himself realizes. Rey can shoot Force lightning like a Sith because the Jedi–Sith binary is a problematic illusion. So Han's cynical remark—which, remember, he makes in the lounge part of the *Falcon* while they are in the lounge on the endless Saturday afternoon of hyperspace—is the first truly spiritual thing in the film!

Telekinesis, a powerful feature of the force, implies how things can travel faster than light. I can move this glass over there without walking over and grabbing it. I can move it as soon as I want. In other words, I can move it faster than light. The *Falcon* making hyperspace is the truly utopian version of the energy-substance that the religiosity of the Jedi hides and turns into an invaluable commodity, creating class division like how Tibetan society recognizes reincarnated lamas in the ordinary folk. The *Falcon* isn't a human. The *Falcon* isn't really a person—it's not even a droid. No wonder that for everyone, the *Falcon* just is *Star Wars* and that it can be detached from the *Star Wars* film so readily, that it has its own reality, that anyone can access it. You don't need to be a sacred lama to make hyperspace. You just need to know how to flip some controls. Which brings us to our next and final topic: democracy.

4 ANYONE

Hyperspace is radically democratic. One of the most potent things about how hyperspace circludes a spacecraft is that *this can happen anywhere*. Hyperspace is everywhere, but not in the Newtonian omnipresent way. Hyperspace doesn't have an entry or an exit point. Hyperspace corresponds to the utopian feminist image of the "body" as a being that doesn't have a rigidly defined inside or outside, and that has (as I put it in the introduction) multiple entry and exit ports. The philosopher Luce Irigaray has written extensively about this body.[1]

Hyperspace is "anywhere" in this radical way. Likewise, spacecraft are also radically democratic in that they *can be piloted by anyone*. The philosopher Levi Bryant wrote a book on OOO, about how all beings exist in the same way, that none of them are more real or more special than others. It is called *The Democracy of Objects*.[2] It would be right to say that spacecraft are the objects of democracy.

"I can do this. I can do this." So says Finn, in the gun cockpit in *The Force Awakens*. "I can do this, I can do this," echoes Rey, in the flight cockpit. They are metaphorically saying that they can be characters in *Star Wars*, and that *Star Wars* itself can also "do this"—keep being *Star Wars*. And indirectly Finn and Rey are telling the audience that they can also refresh its ability to "do this"—be part of the action. We audience members are after all the R2D2 that projects the whole thing from our desires. It's a pivotal moment. (Conversely, the whole plot of *The Empire Strikes Back* pivots on whether or not the *Falcon* will function properly. As soon as it does, the film is emotionally "over" and we can await the future, the sequel.)

The repetition—doubled repetition even—is very significant. It means uncertainty. It means that you might *not* be able to "do this." You are reassuring yourself. You are acknowledging that things slip and slide about. Because time is an intrinsic part of things, things aren't really directly present for inspection and mastery. You are coaxing yourself—you are "piloting" yourself through a difficult moment. The first repetition, by Finn, is uncertainty, naked and raw. So is Rey's, but for the audience it's funny, because in a way it's now a kind of reassuring uncertainty. We've just heard that line. We *can* do this. Objects withdraw but that doesn't mean that they're necessarily dark and menacing. It's how they could be friendly or loving or pleasantly familiar.

The *Falcon* was made by a corporation, then it was heavily modified by Lando Calrissian. It's a wonderfully inconsistent being. Small as it is compared with the *Enterprise*, it's full of entry and exit holes, crawl spaces, curving corridors, smuggling compartments, little "O" shaped portholes . . . I note with some gratification that in several shots there are often three of these O-shapes, spelling out "OOO." There's a wide variety of sliders and knobs and wheels and aircraft-like tools. There's a World War II Lancaster Bomber window from *The Dam Busters* (a film about my great uncle Barnes Wallis, incidentally, which Lucas alludes to in several places including this)—or is it the mullioned window from the shop in *Alice Through the Looking Glass*?[3] There's a multispecies holographic chessboard. There's a bedroom. There's a walk-in closet. There's a fuse box on fire. Inconsistent, that's what the *Falcon* is. It's asymmetrical. It's modified with all kinds of kluges and upgrades. You can fix the engine while you're flying—"I *bypassed* the compressor."

The *Falcon* has a front and a back, a top and a bottom (beautifully close together as those are). It's also asymmetrical. The cockpit sticks out on the right-hand side. There's something on the *Falcon* like the traditional nipple-like control center, the kind that you find as the bridge of the *Enterprise*, or the bulb at the top of a fictional flying saucer. But this nipple is situated underneath the craft, where you would find a nipple on a four-legged mammal. And it's the place from where you fire the rear-facing guns. This isn't a command-and-control place. It's where you go if you're not so good at flying, because you need to make yourself useful. You're not looking down and across from above. You're swiveling about below.

In addition to the objects I was saying it resembled in the previous chapter, the *Falcon* is a surfboard, a horseshoe (good luck), or a giant key with two prongs, or perhaps it's a flattened spork. The *Falcon* is something you can tinker with. You can *craft* it—it's like Forky (a modded-out Spork) from *Toy Story 4*.[4] It's a craft in that sense too. In the later 1700s, in European and American culture, a distinction arose between *artist* and *artisan*.[5] This was a class distinction. The artist was higher class than the artisan. The artist made their own stuff, the artisan worked on other people's stuff, like William Blake making engravings for a living. In the English class system of those days, an artisan was upper working class: they weren't as low as an agricultural worker ("peasant") or what was considered even lower, an industrial worker.

But they weren't as "high" as the imaginary class of the "artist"—imaginary because of course, real actual artists might be very poor, extremely so if they didn't do any crafting for anyone else.

When you see the interior of the *Enterprise* or the Death Star, you see people who resemble office workers. They are working in a gigantic open-plan office, often sitting at a desk (a console). They are coded as "middle class"—another kind of imaginary class. There are other workers around—janitors, soldiers, security. Some of them are droids. Droids are definitely the lowest class in *Star Wars*. C-3PO is a butler, who in the late 1700s was considered higher than an artisan. They were upper-class servants. Doctors (also droids in *Star Wars*) and lawyers and scholars were also upper-class servants. The banter between C-3PO on the one hand and Han and Chewbacca on the other reflects some of this distinction.

Han and Chewie are ragamuffin scoundrel criminals from what Marx would have called the lumpenproletariat. It's their greatest break ever to have gotten their hands on the *Falcon*. Now they can rise to the level of artisan. How Chewbacca, who comes from the noble wise Wookiee species, ended up in this predicament is hard to figure out, until we realize in *Solo* that he was captured and tortured by the Empire. There is something magical about how the *Falcon* transforms these figures into artisans. When Han and Chewie tinker with the knobs and controls on the *Falcon*, it's very different from the paper pushers in the Death Star—or even the more progressive-seeming "office workers" on the *Enterprise*. They aren't following rules. They are in a way "playing" the instrument panel like . . . well, an instrument. A musician is another kind of artisan, or *artiste*, a derogatory, feminized term. In a concert hall, the entrance for the musicians is called the "artiste's entrance" and is hidden away somewhere in the parking lot. Han, Chewie, L3 and other *Falcon* pilots smack the controls, fiddle with buttons and levers, they do things one would do to a broken car or to an instrument in a band, they know how to "play" it. They are artisans working with an instrument, not artists making things up out of nothing. As with musicians, their instrument is a really a sort of person, a co-conspirator. We can find another strong example in the back-and-forth musical play between the aliens on the Mother Ship in *Close Encounters* and the Arp synthesizer, played by one of the human welcoming party. While watching and reading about this film obsessively, I was also learning all I could about the radio telescopes at Jodrell Bank in Cambridge. Listening and acting are deeply intertwined.

One must attend to the *Falcon* just as one must listen to one's instrument. Acting in a revolutionary way is not necessarily about artistry, creating something like the patriarchal God in a void. It might be like being in a band. And it's a band anyone can join. Just like a rock band, you can start with nothing and end up becoming world famous—Rey is already a big fan of Han Solo and "fangirls" on him when he boards the *Falcon*. Again, *Star Wars* is about being the audience-chorus, joining in and making fan art, crafting and re-crafting. America (the good parts anyway) is about making strange new things out of the broken bits of Europe lying around. Black Lives Matter is made out of brokenness and so is the blues. It's why America can talk to the world. It's about salvaging junk—in the case of BLM it's about directly confronting the ongoing legacy of slavery, not accepting social space as an office whose furniture (statues of slave-owners, for example) you should respect. You shouldn't respect that furniture. You should tear it up and throw it in the river.

Let's take what we've got—how can we use anything other than the broken, malfunctioning bits of the past lying around?—and make something wonderful out of it. The *Falcon* is an instrument for improvising a revolution. No wonder the libertarians, starting with Reagan, appropriated *Star Wars*. Libertarianism isn't about being truly free. Libertarianism is about being "free" from having to pay taxes to support poor suffering oppressed people, a huge number of them Black, about being "free" to think white supremacist thoughts. *Star Wars* is not a libertarian film. Anyone can fly the *Falcon*, no matter what species, even if they are seemingly not "alive," like L3. (Why? Because she's made of metal and plastic?) Speciesism—the supremacy of the human species—is created out of racism and patriarchy. (This is a rather long argument, so I'm going to have to ask you to believe me, or read a book I wrote called *Humankind*.)[6] If anyone can fly the *Falcon*, it means that the *Falcon* is a place where, in our imaginations at least, racism and patriarchy have been destroyed.

What is particularly intriguing about the *Falcon* is its clownish, comedic quality. Official American uprightness, like Sam the American Eagle in *the Muppet Show*, has to do with resisting the obscene clownish qualities of America, while of course putting them on display. This is very much like how the Enlightenment ideas of "freedom" and "man," as in "We hold these truths to be self-evident . . . " (The Declaration of Independence) are made up of bits of white supremacy and patriarchy. Reaganism was able to steal the *Falcon*, because it looked like it was on their side—the obscene underbelly of America. But it isn't. How do you defeat fascism? By becoming indignant like Sam the American Eagle? But Sam's indignation is made up of obscenity! So, the way to subvert the fascist tendencies in American culture and economics is to steal the *Falcon* back. Stealing the *Falcon* and playing it like an instrument is a nonverbal image of revolution, like those in Sergei Eisenstein's film about the Russian Revolution, *October*.[7]

But unlike in the Russian Revolution, it's not about some artist-like hero seizing the moment (Lenin). It's about people from the criminal class whom Marx sometimes calls the enemy! And they're seizing an instrument and playing it, listening to it, not imposing their will. America is a country that isn't really one, because of slavery. America is (from) the future. If it's something from the past, it can only be a hotbed of fascism. If America is a weird, junky model of future worlds in which there is planet-scale human collective awareness and action, then . . . well the *Falcon* is the utopian quality of America itself. Hint: it's a bird of prey, but it's not fierce and official like an eagle; it's a falcon.

That's why I wrote this book, the noise in the background which I hope is becoming more audible as you read. The whole idea is that American "culture" — which seems to be saying something very compelling and attractive and deeply unconscious to the whole world (it's not just our fault, dear non-American readers, slavery is a worldwide phenomenon . . .)—has some kind of junk in it that we need to stick our hands into, steal and appropriate to create a different world. The official advert for this world is the money-less *Star Trek* and its warp drive, which as I have argued resembles what Blake says about how the future we require is reached through "an increase in sensual enjoyment" in *The Marriage of Heaven and Hell*:

The ancient tradition that the world will be consumed in fire at the end of six thousand years is true, as I have heard from Hell.

For the cherub with his flaming sword is hereby commanded to leave his guard at [the] tree of life, and when he does, the whole creation will be consumed and appear infinite and holy, whereas it now appears finite and corrupt.

This will come to pass by an improvement of sensual enjoyment.

But first the notion that man has a body distinct from his soul is to be expunged; this I shall do by printing in the infernal method by corrosives, which in Hell are salutary and medicinal, melting apparent surfaces away, and displaying the infinite which was hid.

If the doors of perception were cleansed everything would appear to man as it is, infinite.

For man has closed himself up, till he sees all things through narrow chinks of his cavern.[8]

The "consuming" fire looks like Blake's own way of imagining hyperspace. We imagine it as liquid and bluish, but what's not to like about how Blake figures the "end of the world" as movement, as a flickering fire? Blake's carnivalesque writing turns the idea of hell as punishment for enjoyment into enjoyment itself: it's only hell when a priest tells you not to enjoy yourself. He turns infinity from continuing forever (omnipresence) into an image of dissolving the philosophical and religious binary of soul and body. This binary, which is also in the subject versus object binary, is always about masters and slaves.[9] Blake's future has to do with an end to slavery.

But *Star Trek* is ultimately very sincere and Sam-the-American-Eagle-like. How it feels, how to do it, how to achieve the improvement in enjoyment, how to "make hyperspace"—that's what *Star Wars* is all about. The Jedi are an irritating sideshow to all this—they are how it all goes wrong, by becoming religious. Sam the American Eagle wrote *Star Trek*. Gene Roddenberry's heart was in the right place—the basic rule of *Star Trek* is that any race, class or gender should be on the show. The first televised interracial kiss (between Uhura and Kirk) was on that show.[10] One of the ways in which enjoyment must increase in the future is an ecological one. Humans must get used to appreciating, enhancing, and defending the pleasures of nonhumans—that's a pretty good definition of a more ecologically attuned coexistence. One of the nonhumans that space pilots interact with are their spacecraft.

Doctor Who's TARDIS is definitely a person with whom one must collaborate rather than a tool to be pushed around. In one episode, written by Neil Gaiman, the TARDIS becomes a woman who asks whether all people are like how she feels, bigger on the inside?[11] The "anyone" quality of the *Falcon* makes it into an interesting version of the TARDIS. Spacecraft are nonhuman beings with whom humans readily interact. But there is one actual, terrestrial nonhuman being that features regularly in science fiction as a spacecraft: the whale. Perhaps this derives from the story of Jonah and the Whale, and more immediately, from the huge outpouring of ecological affection and political action to stop whaling from the 1970s on. Whales as terrestrial aliens, spacecraft-like in size.

Whales fly in space in *Doctor Who*, *Star Wars Rebels* and *Star Trek: The Voyage Home*.[12] Dolphins escape from Earth before it is destroyed in *The Hitchhiker's*

Guide to the Galaxy, having tried unsuccessfully to communicate Earth's imminent destruction by doing tricks at aquariums.[13] A nuclear missile is turned into a sperm whale by the *Heart of Gold*'s infinite improbability drive, and we hear its poignant internal monologue as it hurtles toward the ground ("Who am I? What's my purpose in life?"). The melancholy of this example reflects how in general, sci-fi codes whales as beings with whom humanoids ought to cooperate, as fellow "subjects" often cruelly feared, or treated as "objects." In *Star Wars Rebels* the whales can travel through hyperspace and communicate telepathically.

The environments in which we discover the *Falcon* are also radically democratic, on the margins of acceptable society, the Galaxy's garbage patches. Washed-up has-beens, gamblers, criminals, spies . . . perhaps the most *Falcon*-ready space is the holy city on the planet Jedha in *Rogue One* (2016). But we might detect an echo of the *Falcon* here as the main characters' U-Wing makes another improbable jump to hyperspace—this time from within the circluding curl of a wave of rocks. In the subsequent scene the proto-rebels steal an imperial cargo vessel, which creates a strange resonance between it and the shuttle that is the *Falcon*, and the vessel of the Emperor in the prequels and *The Return of the Jedi* (1983). The stolen imperial vessel becomes the primordial model for the kind of craft that the *Falcon* is.

Often the primordial thing turns up afterwards, in the chronology. Think about a virus. There couldn't be viruses before there were single celled organisms such as bacteria. But viruses say something true about life. DNA and RNA can slip out of a necessarily permeable cell wall. Life involves permeable things accidentally swallowing things, happening upon them like Rey and Finn happening upon the *Falcon*. Is it garbage or is it good for you? That's the whole issue with symbiosis. You can't know in advance. There you are, blobbing through the ocean. Plop! Something goes in. Did you just swallow poison? Is this romantic partner the one who is going to ruin your life? You can't know in advance. If you think you can, then you end up with walls and you can't have life. Life implies dirt, disgust, uncertainty. Beauty depends on these things—it's not the opposite of them.

Perhaps this is why the *Falcon* is a radically, even subversively, democratic spacecraft. *Anyone can fly it.* (Now that there is a Falcon Disney ride, this is quite literally true.) All the right decisions are made in advance in the *Star Trek* franchise, as if the current states of play regarding race and gender in the 1960s, 1970s, 1980s, and 1990s were incorporated in the flight manifests of the 2300s. There is already a paradox here, created by the good intention of the impulse. By 2300, these protocols will be outdated. With any luck, human society, may have become much, much more progressive than we can now imagine. So, from a strict point of view, the Starship *Enterprise* has an atavistic crew selection protocol, as if the airliners of today were staffed according to the race and gender rules of the 1600s. (And not coincidentally, modern airliners *can* feel like this, from time to time.)

As I mentioned in the Introduction, *Star Wars* has always been *Muppet Star Wars*, which is why the actual Muppets got involved in the second film. It was as if the Muppets had already been involved, in all the extraordinary creatures we find at Mos Eisley spaceport. The Muppets are in some sense the posthuman beings we have been waiting for. The planetary anthem of the future I want to live in should be "The Rainbow Connection," and surely the planet-scale "America the Beautiful" or "Land of Hope and Glory" should be "We Are All Earthlings."[14]

The true resolution of the *Star Wars* tension would be if one of the characters realized they were actually in an episode of *The Muppet Show*. They are after all almost constantly surrounded by Muppets. It is as if each character in *Star Wars* is a "confused" version of the Buddhas that are the Muppets. For example, Yoda, that perfectly competent green mini-being voiced by

Frank Oz, is surely secretly Gonzo, that wonderfully incompetent miniature blue failure voiced by Dave Goelz. Is Darth Vader, with his dog-like muzzle, actually Rowlf the pianist? Wouldn't it be great if he suddenly realized that? Aren't the Sith just Stadler and Waldorf—grey wizened beings who are constantly being negative and drawing attention to failure? The Sith are like cynical movie critics: *look at this screen and witness the failure of your plan*, say the Sith, just like how Stadler and Waldorf say "boo" to almost everything on *The Muppet Show*. The Sith are the audience, not realizing they're the audience. They stand apart. As soon as they realize they're on *The Muppet Show*, all their sins will evaporate. As I've argued, the "sins" of *Star Wars* are all about introducing hierarchy and religiosity and class division and differences between audiences and characters.

I will now cease this line of thought because really, it's way too far out, even for me. But it's an interesting game, isn't it? Can you match characters in *Star Wars* with Muppets? Who is Leia? Surely not Miss Piggy. I'm not sure yet. Who is Luke? I'm thinking Kermit, and I'm thinking that the child Anakin is Robin, Kermit's nephew. Who is Rey? One of the human guests on the Muppet Show?

The *Falcon* really is a millennium falcon, a nonhuman being that announces the possibility of a new age; a tool that isn't ever always just something to manipulate or be manipulated by; a vessel that's also a way of life; a well-equipped weaponized craft that is also a luxurious lounge, dirty and noisy as it can be; a synecdoche for a film about a truly non-hierarchical, spiritual but not religious, pagan way of life that is all about non-mechanical, telekinetic movement and causality. In short, magic.

The whorl of hyperspace features in an Afrofuturist video for "All the Stars" by Kendrick Lamar.[15] Part of the chorus, "All the stars are closer," is very much in the vein of the intimacy with which the lounge of the *Falcon*'s hyperspace radiates. It's about intimacy and it's also about un-alienation.

Space is such a readily available figure for alienation. Stuck on hold with a terrible phone company? You're floating in space near the moons of Jupiter trying to get a paranoid computer to open the pod bay doors (I refer to the memorable scene in *2001*). "All the stars are closer" has to do with humans rediscovering their cosmic superpowers, powers that they often hand over to a divine being or to the state or what have you. As the female singer crescendos into the chorus, the hyperspace whoosh occurs and we suddenly find ourselves in the space of un-alienation—Kalunga, the transdimensional gate between worlds in the Kongo philosophy.

I can't help wanting to end as I started with an image of hyperspace imagined as a habitat—a pond, no less, like Monet's pond at Giverny. The pond in question is the one toward which we descend in the opening shot of *The Muppet Movie*. We find a lily pad, and on this lily pad we see Kermit, singing what I take to be a highly utopian song about a good place that is also no place, just like hyperspace. *Someday we'll find it, the rainbow connection.*

Why are there so many songs about rainbows
And what's on the other side?
...
Someday we'll find it, the rainbow connection
The lovers, the dreamers, and me.[16]

Rainbows, the Bifrost Bridge of Norse mythology, the shimmering, transient rainbow bridge between this world and Asgard, the realm of the gods, hyperspace. The relentlessly gentle quietness of this song is louder than a number of stiffly martial national anthems. And notice how it talks about stargazing, and about sailors, the siren song of utopia, of hyperspace, the stuff that spacecraft launch us into.

The millennium (as in falcon) is hyperspace, that post-apocalyptic moment at which a more just world is established. Can the work ever be complete? The future of the future is what the word *millennium* is sheltering. Isaiah, the vegetarian atheist poet Percy Shelley who borrowed Isaiah, and many others, visualize in this sci-fi future the lion lying down with the lamb as a consequence of better relations among humans. We get to better interspecies relations through better intraspecies ones. It's what I have been arguing about anti-racism and ending patriarchy. It's something to think about now that we are literally "after the end of the world": *after the end of the idea of world* as a solid bounded horizon for "us" to the exclusion of "them"; and *after the end of the idea of the end of the world*, knowing that things will carry on, in whatever state we help to make them. America never was a country, never did achieve escape velocity from slavery and property. But America could still be a junkyard with a few hyperdrives lying around, looking like garbage. America never was an eagle. It might however be a falcon. I'm glad, Yeats, that the falcon can't hear the falconer as it turns and turns in the widening gyre. Good riddance.

NOTES

Introduction

1 Edmund Husserl, "Prolegomena to Pure Logic," *Logical Investigations Volume I*, J. N. Findlay, translator (London: Routledge, 2001).

2 Jacques Derrida, *Speech and Phenomena: And Other Essays on Husserl's Theory of Signs (Studies in Phenomenology and Existential Philosophy)* (Evanston: Northwestern University Press, 1973).

3 Martin Heidegger, *Being and Time*, tr. Joan Stambaugh (Albany, NY: State University of New York Press, 1996), 17–23.

4 Robert Farris Thompson, *Flash of the Spirit: African and Afro-American Art and Philosophy* (New York: Vintage, 1984), 109, 189.

5 Numerous directors, *Doctor Who* (BBC, 1963–).

6 *Close Encounters of the Third Kind* (1977). Carl Sagan, Ann Druyan, and Steven Soter, *Cosmos: A Personal Voyage* (PBS, 1980); Carl Sagan, *Cosmos* (New York: Random House, 1980).

7 Pink Floyd, *Animals* (EMI, 1977).

8 Amy Hollywood, *Sensible Ecstasy: Mysticism, Sexual Difference, and the Demands of History* (Chicago: University of Chicago Press, 2002).

9 Lisa Yaszek, "Feminism," *The Oxford Handbook of Science Fiction*, ed. Rob Latham (Oxford: Oxford University Press, 2014).

10 Luce Irigaray, *This Sex Which Is Not One*, tr. Catherine Porter and Carolyn Burke (Ithaca: Cornell University Press, 1985), 106–18.

11 Nicholas Roeg, director, *The Man Who Fell to Earth* (British Lion Films, 1976).

12 This is a concept called *subscendence* that I develop at length in my *Humankind* (Verso, 2017), 101–20.

13 David Lynch, director, *Wild at Heart* (The Samuel Goldwyn Company, PolyGram Filmed Entertainment, Propaganda Films, 1990).

14 Ronald Reagan, "Remarks at the Annual Convention of the National Association of Evangelicals," speech (Orlando, Florida: March 8, 1983).

15 James Thomson (lyrics) and Thomas Arne (music), "Rule Britannia." James Thomson, *Complete Poetical Works of James Thomson*, ed. Robertson, J. L. (Oxford: Clarendon Press, 1908).

16 Margaret Thatcher, interview, *Women's Own* (September 23, 1987), https://www.margaretthatcher .org/document/106689, accessed November 24, 2020.

17 Numerous directors, "Pigs in Space," *The Muppet Show* (Associated Television, Henson Associates, ITC, ITC Entertainment, 1977–81).

18 Timothy Hill, director, *Muppets from Space* (Columbia Pictures, 1999).

19 Peter Sloterdijk, *Critique of Cynical Reason* (Minneapolis: University of Minnesota Press, 1987).

20 Rodney Bennett, director, *Doctor Who: The Ark in Space* (BBC, 1975).

21 Paul W. S. Anderson, director, *Event Horizon* (Paramount, 1997).

22 James Strong, director, *Doctor Who: Voyage of the Damned* (BBC, 2007).

23 Terry Jones, director, *Monty Python's Life of Brian* (Cinema International Corporation, 1979).

24 Samuel Taylor Coleridge, *The Rime of the Ancient Mariner*, in Coleridge's *Poetry and Prose*, ed. Nicholas Halmi, Paul Magnuson and Raimonda Modiano (New York: Norton, 2004), 5.18–26.

25 Robert Zemeckis, director, *Contact* (Warner Brothers, 1997).

26 Carl Sagan, "Pale Blue Dot" (1989), https://www.youtube.com/watch?v=wupToqz1e2g&t=47s, accessed November 24, 2020; *Pale Blue Dot: A Vision of the Human Future in Space* (New York: Random House, 1994).

27 Percy Shelley, *Queen Mab* 2.429. *Shelley: Poetical Works*, ed. Thomas Hutchinson (London and New York: Oxford University Press, 1970).

Chapter 1

1　Nicholas Royle's book *Veering: A Theory of Literature* (Edinburgh; Edinburgh University Press, 2011) says more about this in a more powerful and subtle way than I can possibly imagine.

2　Denise Ferreira da Silva, *Toward a Global Idea of Race* (Minneapolis: University of Minnesota Press, 2007).

3　Mary Douglas, *Purity and Danger: An Analysis of Concepts of Pollution and Taboo* (London and Boston: Routledge and Kegan Paul, 1980).

4　Graham Harman, *Object-Oriented Ontology: A New Theory of Everything* (London: Penguin, 2018).

Chapter 2

1　Marcel Mauss, *The Gift: The Form and Reason for Exchange in Archaic Societies*, tr. Halls, W. D. (New York and London: W.W. Norton, 1990); Scott Shershow, *The Work and the Gift* (Chicago and London: University of Chicago Press, 2005).

2　Terry Nation, *Blake's 7* (BBC, 1978–81).

3　Douglas Adams, *The Hitchhiker's Guide to the Galaxy* (BBC Radio 4, 1978–2018).

4　In chapter one of *What Is Property? An Inquiry into the Principle of Right and of Government* (1840), "https://www.marxists.org/reference/subject/economics/proudhon/property/ch01.htm," accessed November 27, 2020.

Chapter 3

1 Mel Brooks, director, *Spaceballs* (MGM, 1987).

2 Eddie Izzard, "Death Star Canteen," *Circle* (Vision Video, 2000). Copyright restricts viewing in the USA, so here is a most amusing LEGO version: "https://www.youtube.com/watch?v=Sv5iEK-IEzw," accessed November 23, 2020.

3 *Oxford English Dictionary*, "hyper-," prefix. oed.com, accessed October 10, 2020.

4 Albert Einstein, *Relativity: The Special and the General Theory* (London: Penguin, 2006).

5 Immanuel Kant, *Critique of Judgment: Including the First Introduction*, tr. Werner Pluhar (Indianapolis: Hackett, 1987), 519–25.

6 *Oxford English Dictionary*, "whorl," *n.*, oed.com, accessed April 4, 2021.

7 Timothy Morton, *Hyperobjects: Philosophy and Ecology after the End of the World* (Minneapolis: University of Minnesota Press, 2013).

8 Sarah Brightman and Hot Gossip, "I Lost My Heart to a Starship Trooper," single (Ariola Hansa, 1978).

9 Alan Rabold, private communication.

10 William Blake, *The Marriage of Heaven and Hell*, plates 25–27, in David Erdman, ed., *The Complete Poetry and Prose of William Blake*, ed. David V. Erdman (New York: Doubleday, 1988).

11 Bernard Newnham, "Norman Taylor's story of Dr Who," *The Tech-ops History Site*, November 28, 2010.

12 Chögyam Trungpa, *The Sadhana of Mahamudra*, *The Collected Works of Chögyam Trungpa*, ed. Carolyn Rose Gimian (Boston: Shambhala, 2003).

13 Bini Adamcszak, "On Circlusion," Mask, July 2016. "http://www.maskmagazine.com/the-mommy-issue/sex/circlusion," accessed November 11, 2020.

14 Oxford English Dictionary, "swivel," v., "swive," v., "https://www.oed.com," accessed November 27, 2020.

15 *Oxford English Dictionary*, "weird," n. 1.a., 1.b., 2.a. "https://www.oed.com," accessed November 27, 2020.

16 Kant, *Judgment*, 519–25.

17 William Blake, "Auguries of Innocence," *The Complete Poetry and Prose of William Blake*, ed. David V. Erdman (New York: Doubleday, 1988).

18 Timothy Morton, *Hyperobjects: Philosophy and Ecology after the End of the World* (Minneapolis: University of Minnesota Press, 2013), 79.

19 Plato, *Republic* 7.514a–517a. *The Republic*, tr. P. Shorey, P , 2 vols. (Cambridge, MA: Harvard University Press, 1956).

20 Andrei Tarkovsky, director, *Stalker* (Mosfilm, 1979).

21 Friedrich Nietzsche, *The Birth of Tragedy: Out of the Spirit of Music*, tr. Shaun Whiteside, ed. Michael Tanner (London, Penguin, 1994).

22 Robert Bellah, Religion in Human Evolution: From the Paleolithic to the Axial Age (Cambridge: Belknap, 2011).

23 TIFF Originals, "Douglas Trumbull Master Class: Higher Learning," YouTube video, 1:37:42, December 13, 2012.

24 Con Pederson, director, *To the Moon and Beyond*, short (Graphic Films, 1964).

25 Sigmund Freud, *Beyond the Pleasure Principle*, in *Beyond the Pleasure Principle and Other Writings*, tr. John Reddick, intro. Mark Edmundson (London: Penguin, 2003), 43–102.

26 Khalil Joseph, director, *Until the Quiet Comes*, film for Flying Lotus (What Matter Most, Warp Films, 2013).

27 Alfred Hitchcock, director, *Vertigo* (Paramount, 1958).

28 Chögyam Trungpa said this at Tail of the Tiger, Vermont, early 1970s (attribution unknown).

29 The Beatles, "A Day in the Life," Sgt. Pepper's Lonely Hearts Club Band (Parlophone, 1967).

30 Theodor Adorno, "Sur l'Eau," *Minima Moralia: Reflections from Damaged Life*, trans. E. F. N. Jephcott (New York: Verso, 1978), 155–7 (157).

Chapter 4

1   Luce Irigaray, *This Sex Which Is Not One*, tr. Catherine Porter and Carolyn Burke (Ithaca: Cornell University Press, 1985).

2   Levi Bryant, *The Democracy of Objects* (Ann Arbor: Open Humanities Press, 2011).

3   Michael Anderson, director, *The Dam Busters* (Associated British Pathé, 1955). Lewis Carroll, *Alice Through the Looking Glass* in *The Annotated Alice: The Definitive Edition*, ed. Martin Gardner (New York: Norton, 2000).

4   Josh Cooley, director, *Toy Story 4* (Walt Disney Pictures, Pixar Animation Studios, 2019).

5   Raymond Williams, *Culture and Society: Coleridge to Orwell* (London: The Hogarth Press, 1987); *Keywords: A Vocabulary of Culture and Society* (London: Fontana Press, 1988).

6   Timothy Morton, *Humankind: Solidarity with Nonhuman People* (Verso, 2017).

7   Sergei Eisenstein, *October* (Sovkino, 1928).

8   William Blake, *The Marriage of Heaven and Hell*, plates 25–7, in *The Complete Poetry and Prose of William Blake*, ed. David V. Erdman (New York: Doubleday, 1988).

9   Denise Ferreira Da Silva, *Toward a Global Idea of Race* (Minneapolis: University of Minnesota Press, 2007).

10 David Alexander, director, "Plato's Stepchildren," *Star Trek* 3:10 (Paramount, 1968).

11 Neil Gaiman, "The Doctor's Wife," *Doctor Who* series 6, episode 4 (BBC, 2011).

12 Leonard Nimoy, director, *Star Trek IV: The Voyage Home* (Paramount Pictures, 1986). Mel Zwyer, director, *Star Wars Rebels: The Call* (Disney, 2016). Andrew Gunn, director, *Doctor Who: The Beast Below* (BBC, 2010).

13 Douglas Adams, *The Hitchhiker's Guide to the Galaxy* (BBC Radio 4, 1978–2018).

14 Kermit the Frog and Jim Hensen, "Rainbow Connection," single (A&M Studios, 1979, LP); A Boy and the Anything Muppet Animals, "We Are All Earthlings," Track #1 on *We Are All Earthlings* (Golden Music, 1993, cassette).

15 Kendrick Lamar, "All the Stars" (Top Dawg, 2018).

16 Kermit the Frog, "The Rainbow Connection," in James Frawley, director, *The Muppet Movie* (Henson Associates, ITC Films, 1979), 1–8.

LIST OF FILMS AND OTHER MEDIA

Abrams, J. J., director, *Star Trek* (Paramount Pictures, Skydance Media, Bad Robot, 2009).

Abrams, J. J., director, *Star Trek: Into Darkness* (Paramount Pictures, Skydance Media, Bad Robot, 2013).

Abrams, J. J., director, *Star Wars: The Force Awakens* (Twentieth-Century Fox, 2015).

Edwards, Gareth, director, *Rogue One: A Star Wars Story* (Walt Disney Studios Motion Pictures, Lucasfilm Ltd., 2016).

Frawley, James, director, *The Muppet Movie* (Henson Associates, ITC Films, 1979).

Howard, Ron, director, *Solo: A Star Wars Story* (Lucasfilm, 2018).

Kershner, Irvin, director, *Star Wars V: The Empire Strikes Back* (Twentieth-Century Fox, 1980).

Kubrick, Stanley, director, *2001: A Space Odyssey* (MGM, 1968).

Lucas, George, Director, *Star Wars IV: A New Hope* (Twentieth-Century Fox, 1977).

Marquand, Richard, director, *Return of the Jedi* (20th Century Fox, 1983).

Nolan, Christopher, director, *Interstellar* (Paramount, 2014).

Spielberg, Steven, director, *Close Encounters of the Third Kind* (Columbia Pictures, EMI Films, 1977).

Tarkovsky, Andrei, director, *Solaris* (Mosfilm, 1972).

Trumbull, Douglas, director, *Silent Running* (Universal Pictures, 1972).

Wise, Robert, director, *Star Trek: The Motion Picture* (Paramount Pictures, 1979).

INDEX